Musa Güner

History of Zaza People

Kulturwissenschaft / Cultural Studies
Estudios Culturales / Études Culturelles

Band / Volume 74

LIT

Musa Güner

History of Zaza People

The Origin of the Zaza People based on DNA

LIT

Cover image: Diagram – Autosomal DNA of Population in Turkey (according to MyHeritageDNA)

All illustrations (pictures, maps, diagrams and drawings) are by the author.

Bibliographic information published by the Deutsche Nationalbibliothek
The Deutsche Nationalbibliothek lists this publication in the Deutsche Nationalbibliografie; detailed bibliographic data are available on the Internet at https://dnb.dnb.de.

ISBN 978-3-643-91719-5 (pb)
ISBN 978-3-643-96719-0 (PDF)

Contact:
Fresnostr. 2 D-48159 Münster
Tel. +49 (0) 2 51-62 03 20 Fax +49 (0) 2 51-23 19 72
e-Mail: lit@lit-verlag.de https://www.lit-verlag.de
Distribution:
In the UK: Global Book Marketing, e-mail: mo@centralbooks.com

CONTENTS

PREFACE

The strongest driving force behind the research presented in this paper is a desire to document the history of the Zaza people. All people are bound by their own environment. No one manages to take the pieces of justice between continents and cast their stories together into one complete whole. The historian must begin by borrowing from DNA scientists, archaeologists, geologists, anthropologists and linguists. It has long been considered that the historian should be content with traces found in writing, so a story actually begins with the art of writing and the preservation of written documents. What lies ahead of the cuneiform and the hieroglyphics, the historian should leave to the archaeologists to reconstruct and interpret. An illustrative real-life example, that provides solid guidance on how to read and interpret the most important sources.

Since the 2000s, DNA technology has undergone remarkable development. It has revolutionized the work of forensic technicians, as well as biological and medical research and in recent years has begun to contribute to new knowledge in archaeology and history. Even the genealogists have begun to use DNA as a tool. With the help of small variations in the DNA sequence, it is now possible to find cousins, siblings and even relatives who lived very long ago. DNA testing can help discover fascinating family histories and solve the most difficult genealogical mysteries. Many tantalizing challenges and interesting discoveries await as one enters the exciting world of DNA genealogy. People are guaranteed to find unknown relatives, who can develop into nice new genealogy friends.

In this book, I have tried to present a DNA diagnosis of the Zaza people. The intention is to illustrate the vast diversity and range of DNA work. DNA is the tree of the unwritten folk groups. Effective isolation has given the Zaza people their own DNA. It shows that exclusivity and endogamy in the heartland have made the DNA of the Zazas a distinctive species. DNA, languages, migrations, religious environments, living habits, agricultures and cultural activities are all pieces of the puzzle that contributes to knowledge of the ancient Zaza people. The reasoning behind this work is to utilize a creative way to research our ancient ancestry.

Taking the initiative to delve into the way of the Zaza people through history is a staggering thought. The Zazas and the Alevis in Turkey lived in a tribal society and applied strict **endogamy**. In fact, we Zazas are inter-married and it is enough for the Zaza people to go back 2,500 years to find a common ancestor. The key to finding the relationship between DNA matches is to compare home regions, the heartland "Zazadiyar". The phenomenon is common in regions where the genera are particularly intermarried and closely related to each other, such as in the provinces of Tunceli, Erzincan, Bingöl and Muş. DNA tests show with fascinating clarity the kinship that ties Zaza tribes together. More than 11,500 potential DNA relatives have appeared. DNA tests provide a calculation of how close we are to relatives. The challenge for me was finding exactly how we are related.

From as early as 1962, I had decided to dedicate my life to finding out my origins. I learned about my ancestry after trying six DNA tests. The results of which can vary widely depending on which company one uses. My interest is not just to understand where my genetic ancestry comes from, but also to see how closely all Zazas are linked to each other. My user experience and results were quite different for each company. The overall results were generic and the ethnicity categories were too broad.

Musa Güner, Malmö 2025

PREHISTORY OF THE ZAZA

On the high plains of Eastern Anatolia, where the conquering expeditions of high cultures never reached, the uncircumscribed Zaza people lived. Most mountain villages were cut off from the outside world, but the low plains have always been the scene of many invasions by different empires.

While the region does not have a clear history, it does have a wealth of parallel stories. The real history of the Zaza people is inaccessible to posterity and their life in Eastern Anatolia is still relatively unknown. For this reason, they have not received a place in Turkish history, because the people have never had their own land.

The Zaza heartland extends to the Upper Euphrates, north-west of Lake Van and includes the provinces of Tunceli, Bingöl and Muş. This means that the oldest history of the Zaza people can be found at the Euphrates and Tigris source flows. This ancient people have lived in this region for as long as anyone knows. They have no history in the Hurrian, Hittite and Sumerian sense. As the Zaza people have never had a written language, they have left no written literature, only some stone engraved art (stone art). Their language survived through spoken language.

Posterity therefore knows little about their language and their political life. This ancient people lived during the centuries prior to writing systems and posterity was not interested in their names. Since this is a question of transcription from foreign alphabets, both parts should be proper. Finally, the sources also do not disclose the relationship of the Zazas with other peoples in the region. There is not much that the posterity knows about the oldest destiny of these people. Life in the heartland, “Zazadiyar”, was entirely anonymous. The frequent word “Diyar” in the Zazaic language is the expression of country. For example, “Diyar-î Bekir” is Northern Mesopotamia and “Diyar-î Rabya” is Eastern Anatolia.

There are no ancient remains from ancient cultures, nor are there any historical landmarks. However, **Göbekli Tepe** (9600 BC) and **Çayönü** (8630 BC) in Eastern Anatolia were first discovered in 1963, which were at the heart of an emerging cult.

Adobe (sun-dried brick) **“kerpiç”** is the building material used in much of the poorly forested Eastern Anatolia. Due to its low resistance to weather and wind and the need for constant new house construction on the remains of the old ones, there are large **“höyük”** formations that mark the place of the old settlements. The building material is of adobe with chaff and mortar that consists of muddy soil. Chaff (straw material serving as reinforcement) is used when the clay tile is dried to prevent it from cracking and to act as a binder. The prehistoric works of the Tower of Babel and Hanging Gardens of Babylon in Mesopotamia and the Mastaba Pyramids in Egypt were similarly built.

Figure 1. Stone Art 2000 BC – Eastern Anatolia

Anatolia was the land of the ancient Hittites. Early inhabitants probably entered the region from the north and west and in the south and east were the early civilizations of the Fertile Crescent. As a bridge between Europe and Asia, this region was at the intersection of trade routes, armies and empires and saw the Sumerians, Persians, Greeks and Romans all come and go. The heartland itself of the Zazas was invaded by ancient migrant peoples, including the Hurrians, Medes, Urartians, Cimmerians and Scythians from the Caucasian regions. However, these groups have been perceived as invasive and oppressive. Throughout history, the local population has successfully fought against invaders to preserve their unique character. They have also been largely unaffected by modernity and are associated with an agricultural lifestyle.

The fate of Eastern Anatolia during 1800 BC is associated with the large settlements in the region between Hurrians and Hittites. Each city or city state lived its own life, although in constant peaceful and warlike interaction with others. During the Sumerian era, contact between neighbouring peoples and advanced cultures in the north, east and west was quite slow. The Sumerians disappear from history around 1500 BC, but their language lives on as literature and the sacred language of religion.

Figure 2. Göbekli Tepe 9600 BC – Eastern Anatolia

GEOGRAPHY OF THE HEARTLAND

Eastern Anatolia is largely dominated by a mountainous landscape. The geography is divided into two terrain conditions, high tablelands and lowland plains. The Zaza heartland region consists of 70% mountains, 25% plateaus and 5% plains. Despite its small area, Zaza-diyar has one of the most varied topographies in Eastern Anatolia. The rock formations form an inaccessible alpine landscape with rugged cliffs and some vegetation. High mountains and deep valleys dominate the heartland. The mountains are bare and with an altitude of 2,500-3,500 metres above sea level and the valleys are treeless with embedded alluvial deposits.

High mountains and mountain ranges, up to 3,400 metres like the Mercan Dağları and deep valleys are characteristics of the region of the Northern Tunceli. The whole area is mountainous with small forests or bushes, but lot of water springs and larger or smaller creeks. The northern border is marked by the headwater of the Euphrates River. Pülümür Çayı and others flow into the many ribbons of the Euphrates. The Keban Barajı with its huge dammed up water reservoirs nowadays forms the southern border of the northern homeland. Main cities are Densim/Tunceli, Mazgirt, Çemişgezek, Ovacık, Pülümür, Nazımiye, Hozat, Pertek, Kığı and Varto. The continental climate brings cold winters and short summers to the northern area. In winter, temperatures can be down to -30 °C, whereas summer temperatures do not climb above 28 °C. There is always a fresh breath of air around.

Dersim came under resettlement in 1938, following a period of uprising against national forces, and was settled under the so called "Emergency Rule" **(Olağanüstü Hal)** in 1994. In both periods many villages were emptied and burnt down by the national forces. In between, after 1948 when the ban on settlement was bred, many Zaza people went back to their homeland area and rebuilt their villages.

Most people are poor, undernourished or malnourished. The houses they live in are more or less on the same level as the ones Medieval Europeans lived in, that is, they had dirt floors, no indoor plumbing, rustic conditions, and the like, although architecturally they are totally different. The houses lack running water and many do not have toilets. Most people use out-houses or nature itself for this purpose. Women come together to do laundry by the rivers a stone's throw from the villages. People live very close to the animals in their daily lives, nearly side by side. Cattle, sheep, goats and donkeys shit around houses day and night throughout the year.

The valley bottoms testify that the swamp fields are impracticable for humans and animals. The road network is not developed in the region. The roads around the mountain villages are impassable and rugged and the plains are roadless. In some villages, it is possible to travel with **ox carts** and in other villages the transport is only carried by **donkeys**. It is not possible to ride an animal in this mountainous region because the valleys are very deep, surrounded by steep precipices and very steep mountains. It is very dangerous to pass through the rapids in the spring when the waters are strong.

Due to the great differences in elevation in the region, combined with missing road networks, many small mountain villages are isolated from the outside world during winter. In Zazadiyar, winter is known for its cold, rain and snowfall that last for more than six months in most regions and large variations between the highlands and the plains. Winters are very cold (down to -30 degrees Celsius) and with 3-4 metres of snow. Many villagers perish in snow avalanches. The summers are very hot and dry (up to 40 degrees Celsius) and may bring no rain for six months.

Figure 3. The Provinces of Turkey

During blizzards, villagers do not dare to go outside and conduct their daily pursuits and many hikers suffer suffocation accidents. The snow supplies the rivers with water. Heavy rain showers wash away all top soil on slopes and from steep cliffs with rocks and gravel carving the mountain gorges. The downstream colloidal sludge material is deposited after a distance of 2,800 kilometres out at Shatt al-Arab in the Persian Gulf. The outlets of the Euphrates and Tigris were separate and lay further north than they do today, as the river sludge having filled in coves and lagoons and pushed the shoreline further out into the Persian Gulf. Glacier water causes many floods in the lowlands. Crossing the violent spring floods is a matter of life and death.

The Munzur mountain ranges are a continuation of the Taurus Mountains to the east and form the border with the Eastern Anatolian provinces of Tunceli and Erzincan. The range extends from Kemaliye in an east-west direction for about 130 kilometres. The middle north-south axis is approximately between the provinces of Erzincan and

Tunceli. Here, the mountains are about 30 kilometres wide. The Mercan mountain ranges are based on the same mountain massif in the east and reach a height of 3,292 metres. The Munzur valleys rise south to Ovacık and merge with the Pülümür valley.

The geological structure of the Munzur Mountains consists mainly of calcareous, ophiolitic and sedimentary rocks, which have been metamorphosed. The construction of the rocks is determined by the attraction of the ophiolitic rocks and their limestone in the north over the sediment in the south. The North Anatolian fault and the Palestinian fault line meet in Erzincan. The North Anatolian fault runs from the Dardanelles to the centre of Iran. The Euphrates River flows through the North Anatolian fault zone. Erzincan is a heavily earthquake-affected area. Due to the most powerful and uninterrupted earthquakes occurring in Erzincan, people are not allowed to build houses higher than three floors. The largest earthquake, measuring **7.9** on the Richter scale, took place in 1939, **33,000** people perished and **4,125** people severely injured. A total of **116,720** buildings were levelled with the ground. Throughout history, Erzincan has undergone devastating major earthquakes during the years 1268, 1458, 1482, 1484, 1784, 1788, 1890, 1930, 1939, 1941, 1967 and 1992.

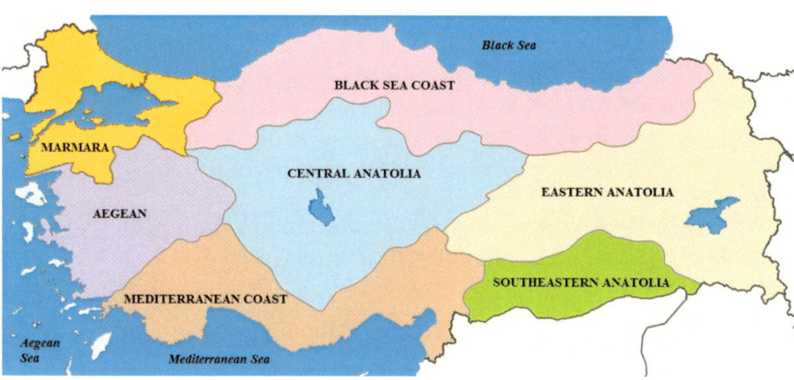

Figure 4. The Regions of Turkey

THE ZAZA LANGUAGE

The Zaza people have an independent language, **Zazaic**, in the Zazaic language ***Zazaki*** or ***Zone Ma***, which belongs to the Iranian language branch of the Indo-European languages, the **Satem language** Indo-Iranian. According to the linguists B. Peter Lerch in 1856 and Ferdinand Justi in 1880, the Zazaic language resembles the extinct (700 BC) Middle Iranian language **Parthian**. This assertion was confirmed in 1932 by the German Iranists Oskar Mann and Karl Hadank, whose investigations resulted in the first extensive scientific study of Zazaic grammar, entitled ***Mundarten der Zaza***. According to the linguist Jost Gippert in his 1996 article "The Historical Development of Zazaic", Zazaic is Old Persian from 4,000 years back. A historical proof of the claim is an Armenian script from **140 AD**, a duplicate of an older script containing Christian prayers in seven languages, one of the languages being "Median". It is written in the language known today as the Northern Zazaic dialect **Kırmançkî**. This work constitutes the oldest preserved Zazaic document.

Zazaic grammar is very weakly inflectional with some agglutinating elements and has cleft ergativity in its morphology. The language is dominated by diphthongs. Sumerian, Elamite, Hittite, Hurrian, Urartian, Zazaic, Persian, Kurdish and Turkish languages have the word order Subject-Object-Verb **(SOV language)**. About 47% of the Earth's languages are SOV languages, including German, Dutch, Classical Latin, Persian, Turkish, Kurdish, Zazaic, Armenian, Japanese, Korean etcetera. A phrase such as "the hunter killed the lion" in Zazaic word order becomes "sayderi şêr kişt" – "sayderi (subject) şêr (object) kişt (predicate)". English is an SVO (Subject-Verb-Object) language, but if it were an SOV language, the phrase "the hunter killed the lion" would become "hunter lion killed".

Zazaic dialects are divided according to the provinces. Zazaic forms three sociolects, in consequence, Northern Zazaic **Kırmançkî**, South Zazaic **Dımılî** and Central Zazaic **Kırdkî**. The largest dialect is Kırmançkî, which makes up around 55% of Zaza speakers. It is estimated that the dialects correspond to Kırdkî about 30% and Dımılî about 15%. The dialects typically differ, among other things, with regard to bending morphology, vocabulary and tone.

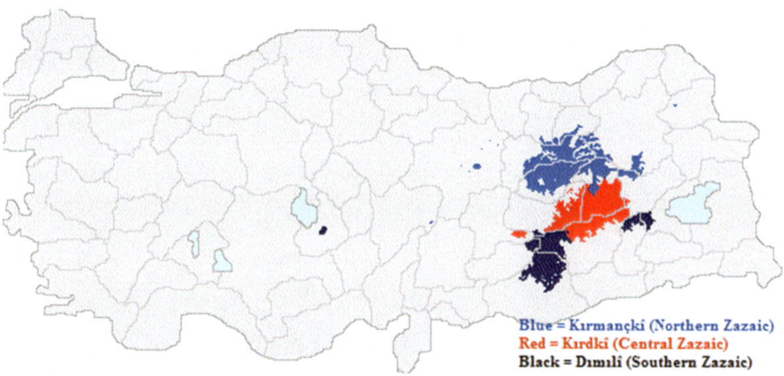

Figure 5. The Zaza Dialects in Turkey

It is a great confusion what the local Zazas themselves wanted to call in their own language. With some caution, one can angle the sociolects in different districts the following: Tunceli (Dersim) "Zazaki, Kırmançkî, Kurmanci, Kırdkî, Here-Were", Erzincan "Kırmançkî", Siverek "Dımılî", Bingöl "Kırmançkî, Kırdaşkî, Here-Were", Muş (Varto) "Zone Ma, Kırdaşkî", Elazığ and Adıyaman "Kırdkî", Diyarbakır and Mardin "Dımılkî", Sivas (Zara, İmranlı, Kangal, Divriği) "Kurmanci", Kelkit "Kırmançkî" and Kemah "Kırmançkî, Kırdaşkî, Kırdkî".

There is a common misconception that Zazaic is a dialect of Kurdish, a closely related language. However, Zazaic differs markedly from Kurdish. Those who believe that Zazaic is a Kurdish dialect have been deceived. The numerals in Zazaic, for example, deviate completely from the pattern found in Kurdish languages, which is one of the most important arguments against kinship with the Kurdish dialects. In Zazaic, the letters **ğ**, **ı** and **u** occur first in a word, which is missing from the Kurdish counter. A completely separate group of languages, **Kırdaşkî**, is spoken by Alevi Kurds in Eastern Anatolia. However, many Zazas in the different provinces have Kırdaşkî as their mother tongue.

Linguistically, Zazaic is most closely related to **Goranî**. The Goranî language homeland is located in the eastern part of the Zagros Mountains in north-western Iran. The difference between Zazaic and Kurdish is the same as that between Swedish and Dutch. The difference between Zazaic and Kırdaşkî is similar to the difference between Swedish and Danish. Kırdaşkî is closely related to and, to a great extent, mutually intelligible with Zazaic.

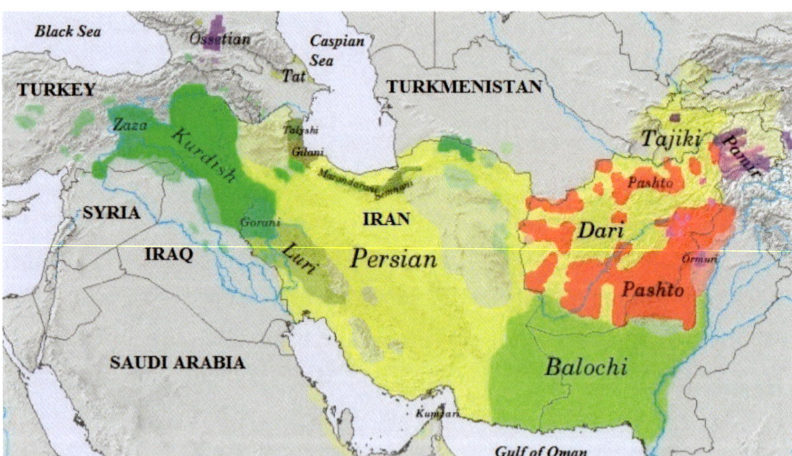

Figure 6. Geographic Distribution of Iranian Languages

The following are examples of numerals **1-10** in the spoken languages of Zazaic and Kurdish:

Zazaic: Zu, Dide, Hire, Câr, Ponc, Ses, Hauet, Heyşt, Neu, Des.
Kurdish: Yek, Du, Sê, Çar, Pênc, Şeş, Heft, Heşt, Ne, Deh.

The following are examples of words to compare between Zazaic and Kurdish:

Zazaic	**Kurdish**	**English**
vatene	gotın	to say/to mention
amaene	hatın	to come/to arrive
şiyaene	çuyın	to go/to depart
uerdene	xwarın	to eat/ to consume
sımıtene	vexwarın	to drink/to absorb
uastene	xwastın	to want/to require
vurayen	guherin	to change/to alter
vac	bêj	to sing/to say
pi, per	bar	father
mae	dê	mother
bıra	brak	brother
uae	xwişk	sister
mordem	mırov	man
ma	em	we
şıma	hun	you
ey	wî	his/him
lıng	pê	foot
zerri	dıl	heart
nake	navokj	stomach
aue	aw	water
sole	xwê	salt
vaş	giya	grass
verg	gurg	wolf
pısing	kıtık	cat
verek	berx	sheep
kerge	mırişk	hen
mircela	muri	ant
mere	mışk	mouse/rat
astare	stêrk	star
veur	berf	snow
raye, rayir	rê	way/road
pil, gal	gir	big/large
veyşan	bırsi	hungry
merg	mırın	dead
neue	nû	new

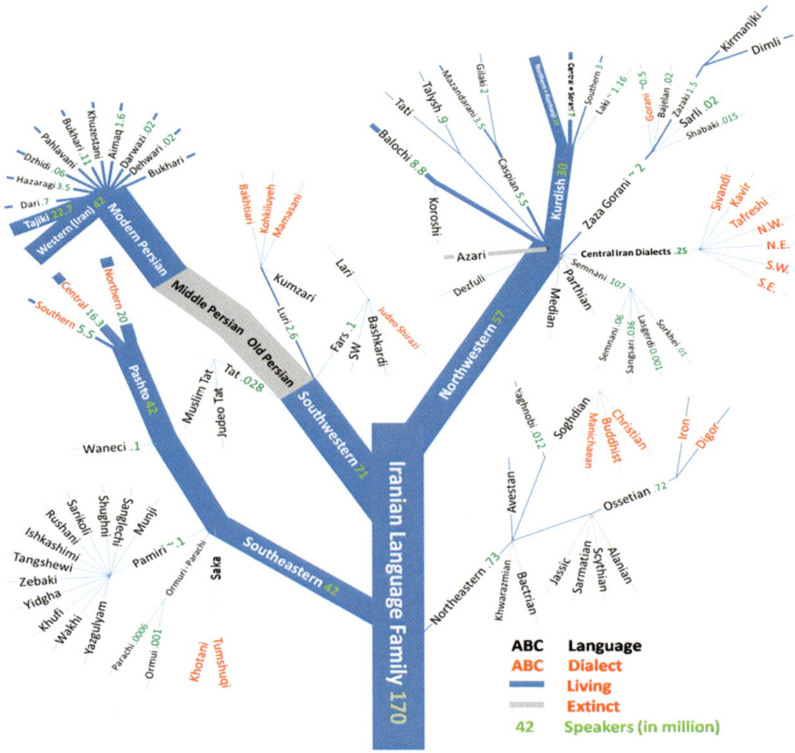

Figure 7. The Iranian Language Family

Zazaic is heavily influenced by Turkish and has a large number of Turkish loanwords which comprise approximately 30% of the vocabulary. However, Erzincan-Zazaic is comprised of about 35% Turkish loanwords. Some loanwords are very heavily modified. Zazaic grammar has been partially influenced by Turkish, including the word-stem-pronounced agglutinating language structure with many suffixes. About **1,000 years** ago, the Zazaic language was completely pure. The multi-millennial language eventually grew into a vernacular language influenced by Turkish and Arabic. In recent times, the Zazas have been given Turkified language in Turkish writing attire. The snowball of change rolled on. It started with the plainsfolk (plain people), while the mountaineers still had "language isolate". The mountaineers were highly dependent on the plainsfolk, who carried the square trade, handcraft, work, etcetera and as a result, the language lost its purity and was forced away in its entirety. Otherwise, not everything else reached the mountain villages.

The origin of the Zazaic vocabulary is assumed to be: **55% Old Persian**, **30% Turkish**, **10% Arabic**, **2% Sumerian**, **2% Kurdish** and **1% Armenian**.

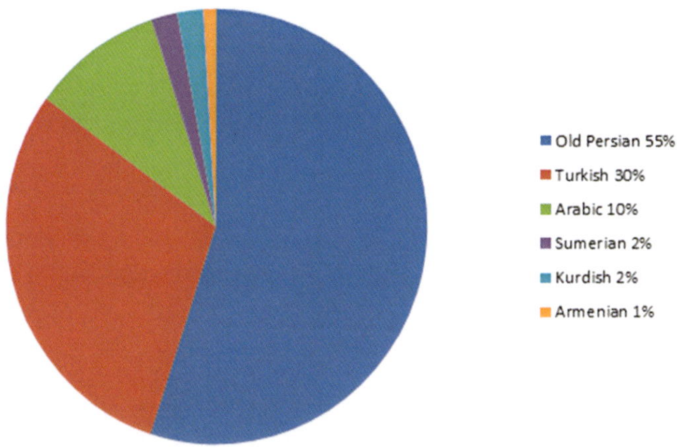

Figure 8. Etymological Distribution of the Zazaic Vocabulary

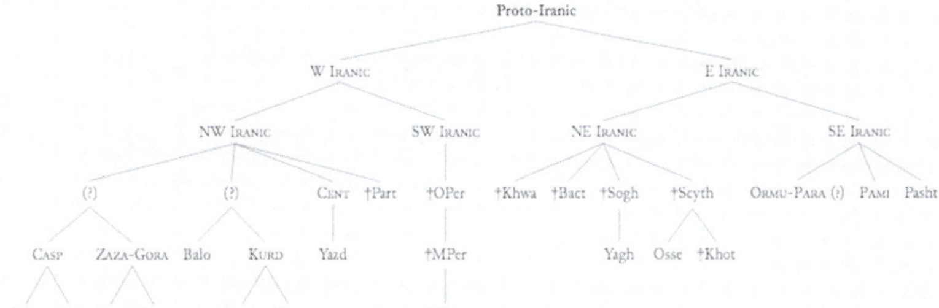

Figure 9. Genetic Division of Iranic Languages

A proverb in the dialect **Kırmançkî** Zazaic reads:

Kam kê aslê ho inkâr keno, akılra hero! (The one who denies its origin is a blockhead).

kam: Old Persian.
kê: Old Persian.
aslê: Ottoman **asıl**, Arabic **asl**. Sumerian agglutinating language structure with ergative suffix **ê**.
ho: Old Persian.
inkâr: Ottoman and Arabic **inkâr**.
keno: Old Persian.
akıl: Ottoman **akıl**, Arabic **akl**.
-ra: Sumerian. Agglutinating structure suffix in dative object.
hero: Old Persian.

Some proverbs imply in **pure Zazaic** from ancient times and read: "Kutık goştê kutıki nêweno" (A dog does not eat dog meat). "Leyrê mori bê zehr nêbeno" (A young snake cannot be without poison). "Moreka lonekerdiye hard de nêmanena" (A glass bead with holes in will not remain on site). "Heş bi pil leyrê bê kayken" (As the bear grows old, he plays fun for his cubs). "Ponc beçiki zu jümini niyê" (Five fingers is not one). All of these words are **Old Persian**, with no trace of ancient Turkish or Arabic loanwords. The grammar is untouched, exception with Sumerian cases, and it contains no foreign rules. To all appearances, the source of these proverbs is unknown.

The following are examples of some words to compare between Sumerian and Zazaic:

Sumerian	Zazaic	English
mar	mar	shovel/dustpan/hoe
taru	tarde	to turn/to twist
lu	melu	man
salamu, sılım	salam	left
giş tug	giştige	box (tree)
gal	gal, pil	big/great/large

arı	arg	river
esık	esige	threshold
ışdı	ışuyde	to sleep
ağıl	ağıl	sheepfold
sur	sur	red
mae	mae, maye	mother
tamkaru	tamkaru	business man/tradesman
arda	ardi	meal/flour/powder
gal	galık	aged/old
bar	bar	load/cargo
meşg	meşke	clothes/dress
si	sich	shade/shadow
agar	ega	field
bar	ber, per	all/every
barna	ban	shed/hovel
duru	vındur, vınduru	to stop
eş	eşan, aşan	to scrape/to scratch
ilu	pil, pilu	great/big/large
kıyı	kıyşt, kışt	shore/beach

The mother tongue in the villages is Zazaic. The men master a little Turkish, which they learn during military service. Most women cannot speak Turkish and in courts, the authorities call the gendarmerie to interpret. The use of Zazaic in Turkey is severely limited. No form of Zazaic language teaching exists. The mother tongue of Zazaic is to a great extent foreign to the new generations born outside Zazadiyar. In the 1930s, no primary schools existed in the small Zazaic towns of Ovacık, Nazımiye, Çarsancak and Pah, each with 2,000-3,000 residents. Public schooling is for five years. The reason for this was that the Zaza children could not speak a word in Turkish and the Turkish authorities refused to pay for public schools. The conflict between the Zaza people and the Turkish education authorities was degenerated. Human brains develop most significantly during early childhood. By the age of five, large parts of the brain's capacity have already been developed. Therefore, without school, children in the Zazaic villages have large knowledge gaps.

THE PEOPLE IN LEGENDS

According to some sources, the word **"Zaza"** is mentioned in Sumerian cuneiform clay tablets (3100 BC). What this refers to may never be known to the posterity. The Behistun inscriptions in Iran, written between 515 BC and 486 BC in Old Persian, Elamite and Babylonian, on the orders of the Achaemenid emperor Darius I, states that the Zaza people lived in a region north-west of Lake Van. According to several undocumented claims, the original home of the Zazas is located in an area near the sources of the Euphrates and Tigris rivers.

Distinctly, historical records such as ***Bundahishn*** (about 500 BC), one of the sacred writings of the Zoroastrian adherents, show the homeland of the Zazas to be in the region around the Euphrates and Tigris source, as they are today. This indicates that the Zazas migrated to Khorasan in Northern Iran, and not the opposite, as previously thought.

The Sumerian word **"Za"** is related to water, especially glacier water. The word **"ab-za"** in Sumerian is an expression of the ocean. The snow supplies the Euphrates and Tigris with water. Water masses, spring floods and floods are linked to the Euphrates and Tigris rivers. The very word Euphrates means "abundance". The Euphrates River in Old Persian is **Ufratu**, in Elamite it is **Ú-ip-ra-tu-iš** and in Sumerian it is **Bu-ra-nu-na**. The Euphrates River has sometimes flooded Sumerian cities and destroyed its monuments over time. The Zaza people's country of origin, "Zazadiyar", is at the source flows of Euphrates and Tigris.

In this respect, if this assumption is correct, it can be asserted that the word "zaza" refers to the residents of these source flows. The Sumerian language is very rich in its duplication of the noun. Examples of this include: **du-du** = the words, **kur-kur** = the mountains, **šu-šu** = the hands, **küren-küren** = the countries and so on. Accordingly, the word "Za-za" can be given its key as "The Flood People" by the country around the source of the Euphrates. Of course, this is a series of guesses and calculations of probability, but it could help these long periods of unknowing less dreary and more tangible.

Figure 10. Behistun-Inscriptions – Kermanshah

According to a Hurrian claim, the Hurrians would have settled in Zazadiyar between 2350 BC and 2150 BC. The Hurrians came from the Caucasian regions. The Greek geographer Klaudios Ptolemaios (85-165 AD) in his work ***Geographike Hyphegesis*** depicted a region south of the Caspian Sea as **"Zazaka"**.

Almost everything we know about Sumerian culture is based on details attained from the excavations of ancient cities have revealed in the form of remains and inscriptions. It is important to history that the high cultures became literary – that is, invented and developed the art of writing. The importance of writing to the communities that created it should not be exaggerated. The writing becomes a language that, at the very least, basically corresponds to what is spoken. The earliest written sources are primitive, difficult to interpret, or completely uninterpretable. Neither at that time nor later were the inscriptions intended to give posterity historical knowledge. The framework of a depiction of these epochs still has to be derived mainly from the archaeology.

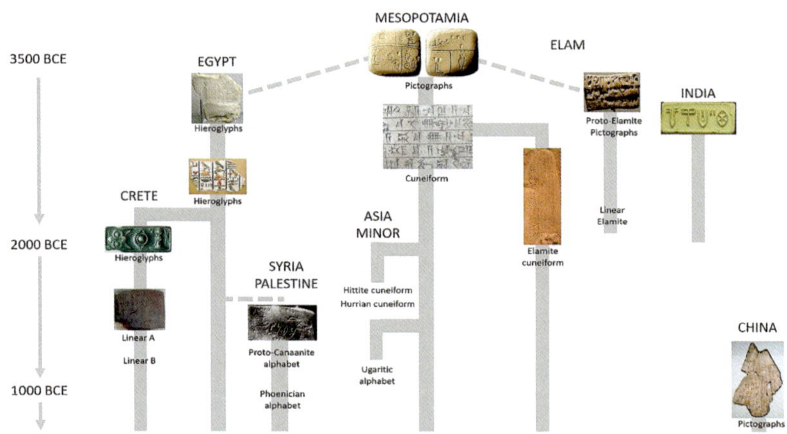

Figure 11. Development of Writing

ESTIMATED POPULATION

Several historical empires have emerged in Turkey as it is one of the oldest continuously inhabited regions in the world. In Turkey, there are 42 significant ethnic groups, in relation to which there are not 42 religions. The population thus belongs to a wide range of different ethnic groups and several of them have strongly differing cultures, traditions and religions.

There are also political and sometimes ethical problems in attempting to categorize the population into ethnicities. Doing so has frequently been used for political purposes, to divide people and exercise power. It is also not entirely clear how an ethnic group should be defined. For example, it is not uncommon for people to marry across the boundaries of ethnicity and then the tradition that is counted relates to the ethnicity of the man and not the woman. People of different ethnicities often live together in peace in Turkey's different cities and villages. However, it can be valuable to have knowledge of some key features of the ethnic groups.

The largest groups are the Turks, Kurds, Zazas, Lazes, Georgians, Arabs etcetera. The figures on the percentage of the population of Turkish ethnicity are sometimes set at 60% and can be as high as 80%. The degree of assimilation to the majority population among different ethnic groups is high. However, practically all Zaza tribes in Eastern Anatolia lived in a tribal society and applied strict endogamy, that is, they married only within their own communities. A high proportion of the intermarriages between the families belong in the provinces of Erzincan, Tunceli, Bingöl, Muş and Elazığ.

The ethnic composition of the Turkish population is highly contested and often difficult to understand as a result of the process of Turkification, the historical interpretation of the sociological definition of ethnicity (for example, the Alevis and Zazas are counted as one group) and the repression of Turkish state power by other ethnicities in the region.

The exact number of Zazas is therefore unknown due to a lack of population studies or detailed censuses. The fact that some Zazas have mixed with other regional ethnicities (mostly Kurds) has also contributed to the uncertainty. Many Zazas live outside their country of origin, ***Zazadiyar***. In addition to widespread oppression and mass evacuation of villages, the poor economic condition of the Zaza regions has forced the local populations to migrate to Turkish metropolitan cities or European countries. Many Zazas live in major Turkish cities such as İstanbul, Ankara and İzmir. Furthermore, Zazas have emigrated to the whole of Europe (mostly to Germany) and even further afield (including America, Canada, Australia and so on).

Figure 12. Distribution of Ethnic Groups in Turkey

Estimates of the number of Zazas are contested, whereby the Zazas' own calculations are consistently higher than those of other sources. According to estimates, there are about two to three million Zazas. According to a survey by the gendarmerie group in 1970, "Reports on Zaza tribes", a total of 350,000 Zaza residents were registered in Eastern Anatolia. More realistically, the number today can be estimated at about 550,000 in Eastern Anatolia, a further 500,000 in the rest of Turkey and around 450,000 outside of Turkey. A rounded figure of **1,500,000** Zazas can therefore be found world-wide. The highest number of Zaza people in Eastern Anatolia is found in the provinces of **Tunceli** (29%), **Erzincan** (15%), **Bingöl** (14%), **Mardin** (7%), **Sivas** (5%), **Diyarbakır** (4%), **Bitlis** (3%), **Elazığ** (2%) and **Gümüşhane** (1%) and in patches propagated in Muş, Erzurum, Urfa, Adıyaman, Van, Bayburt, Kars, Ardahan, Malatya and Maraş. The province of Tunceli is the most populous and consists of about 90% Zazas.

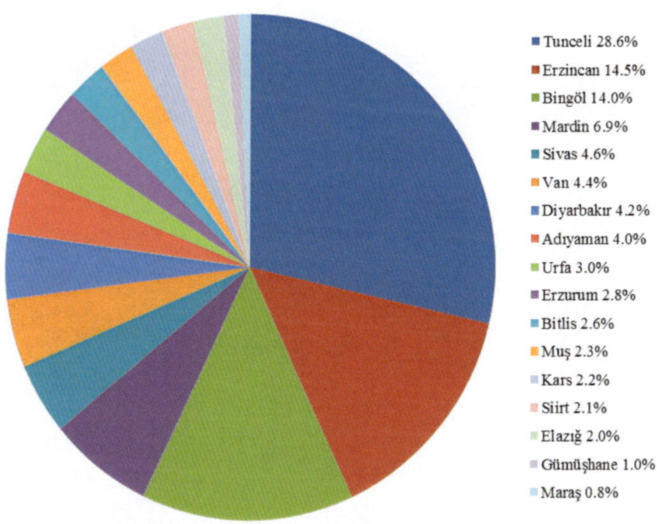

Figure 13. Geographic Distribution of Zazas in Anatolia – Jandarma "Zaza Aşiretler Raporu 1970"

The Zazas have the attractions of modern Turkey and the socioeconomic benefits they offer to other neighbouring minorities that have allowed them to avoid involvement in the Kurdish separatism in the north. In the footsteps of reviving Alevi cultural expressions, the Zazas were able to find their own national self-consciousness in the freer European political climate. Kurdish nationalists have also adopted the method of obstructing cultural and academic activities related to the Zazas. A Kurdish nationalist student group protested against the teaching of Zazaic at Tunceli University. The organization of an international conference on Zazaic by Bingöl University was protested by Kurdish nationalist groups. In some cases, Kurdish nationalists have tried to obstruct activities related to the Zazas by using physical force. Sait Çiya stated that Kamer Özkan was killed by radical Kurdish groups on the pretext that he was a Zaza, that cultural activities organized by Zazas in Europe were subject to physical attacks by Kurdish nationalist groups, and that a Zaza participant was prevented from speaking by Kurdish nationalist groups at a minority group meeting organized by the European Union in İstanbul.

The majority of Zaza people in the provinces of Tunceli, Bingöl, Elazığ and Sivas consider themselves ethnic Kurds and are often described as Zaza Kurds. The exact origin of the name "Kurd" is unclear. In modern times, the word "Kurd" has become a motto. Not least by Abdullah Öcalan, whose father is Armenian and mother is Turkish, who proclaimed himself the Kurdish leader. From the aspect of making Zazas into Kurds, this has national, ethnic, political and social value.

ORIGIN OF THE TRIBES

In the past, the Zazas were basically an ethnic group. However, it was further developed after the Zazas had been exposed to Persian, Roman, Arabic, Mongolian and Turkish intruders after the first part of the past millennium. To this day, the brutality of these intruders has been the subject of conversation. This is an oral tradition that has lived through several tens of generations. Even today, people call these intruders "plunderers, ruthless and brutal occupants". These intruders were so brutal that the Zazas probably had to create their own loyal forces in order to resist them. This happened through organizing in tribes based on kinship bonds. Only strong tribes were able to defend themselves to a certain extent against foreign powers.

The origin of their tribes occurred during the epoch of the Seljuks. It appears that branches of the people arose in the 12th century and about **220 tribes** formed within the Zaza people. Even today, newly formed tribes are a common phenomenon. In a village in Kığı in the 1980s, conflicts between the landowners arose. This resulted in a split of a single tribe into three.

By and large, all the names of the tribes are Turkish. The names are often based on places, local residents, persons, dignities, appearances, occupational branches, patronymics, verbal nouns, compound adjectives and so on. It is rumoured that nine tribes originated in Khorasan in Iran. These would be Koçgiri, Hormek, Hıran, Şadılı, İzol, Karsan, Arılı, Alanlı and Demenan. The tribes have highly varying populations. According to the statistics in 1970's "Reports on

Zaza tribes" for Eastern Anatolia, the largest tribe was the Koçgiri with 21,799 individuals, while the smallest was the Pulanlı with 201 individuals. Broadly speaking, it can be said that the Pulanlı tribe represents a family. Many atrocities took place in Eastern Anatolia. Koçgiri was the most vulnerable tribe and was forced to move to other parts of Turkey. An estimated 60,000 descendants of the Koçgiri can be found around the world today.

In a single year, 1071, Byzantium suffered two disastrous defeats against the Seljuk Turks and lost Asia Minor to the coup. Byzantium's great defeat opencd the doors to Seljuk infiltration. Byzantium steadily lost land to different Turkish states. During the first time, the Seljuks gained control of only the countryside, while the Byzantine cities resisted, however, a little into the 13th century, they attained a firmer state organization. The historical epoch was only slightly larger when the Seljuks took over the political leadership in Eastern Anatolia. The distance in the Seljuk world empire was such that different rulers were trapped in different cultural circles and soon did not have much in common. The Seljuks utilized all elements of the population – Zazas, Kurds, Armenians and Byzantines – for their own interests and preferably against each other. In the war against Byzantium, Zazas participated and contributed heroic efforts. Their warlords carried the titles of **"Ağa"**, **"Bey"** and **"Efendi"**. Many of these Zazas were given very high positions in the Seljuk principalities. Zazaic war heroes were given land in enfeoffment **"ikta"** and the élite were the larger landowners, the most affluent social group. On the whole, then, these dignitaries could contribute to different tribal formations, and from each of these major families a tribe arose. The patterns of the Zaza tribes were rather egalitarian, the families were equal and the central power was weak or non-existent. Most of the affairs were handled within the families and clans themselves, under the leadership of the earthly top families with their clients of poorer and completely landless elements, often joined together in larger fraternities.

Throughout the tribe, the power is delegated downwards so that the tribe is ruled by several subordinate chiefs. However, the tribe can mobilize fully when it is attacked from without or is in conflict with other tribes or is threatened, especially when a member is killed. Then the tribe must seek revenge, otherwise it will lose its honour and place itself in a weak position in relation to competing tribes, and this is a costly situation, something every tribe wants to avoid. Such conditions necessitate that the whole tribe stands together, both to defend its position and to resist the declared enemy. Meanwhile, some tribes cannot manage to activate the lineage principles wholly in situations of conflict. Activation acts more in a fictive manner because, often, it has been seen that some tribes, because of internal discord, could not manage to take revenge. All the same, in many ways, the tribes appear to be decentralized and at the same time rather integrated. Integration is particularly visible and expressive at the segmentary lineage level.

After the tribal formations, there were large-scale migrations. The tribal war made its appearance amongst the feudal lords. The Zaza tribal leaders fought for power among themselves. Due to the constant tribal wars and conflicts, families were divided and there was great of movement and displacement between the villages. The assassination of different tribes/families contributed to lost estates and relocation. Their malicious behaviour often led to a conflict between people, which turned into hatred, humiliation and hostility. These conflicts were based on a number of political goals, and caused by thought, religion and sect infighting. Furthermore, disputes between the descendants of the same race, aboriginal people of the same origin with common interests, continued for many thousands of years. These circumstances instigated a new phase, as the poor Armenians sold their property to the poor Zazas. As the adage states – ***a blind seller finds a blind buyer***.

POPULATION MOVEMENTS

The destiny of the Zaza people has always been overshadowed by constant migrations. The largest may have taken place about 800 BC. Whether relocations have taken place during this century is highly uncertain and open for guesses. There are no concrete details of this migration. The migration routes are understood to have been eastwards to Khorasan in Iran, south-easterly to the Zagros Mountains in north-western Iran and westwards to Cappadocia. In other words, emigration of Zazas from their country of origin to Khorasan and to the Zagros region has been going on since some centuries BC. As noted earlier, historical source writings show that the ***Bundahishn*** (circa 500 BC) locates the homeland of the Zazas around the source of the Euphrates and Tigris. It indicates that the Zazas migrated to Khorasan in Northern Iran, rather than the opposite, as previously thought. These refugees in Khorasan go by the name of Kırmanç (Kurmanç) and the Goranî language was established in the Zagros region. Northern Zazas are called Kırmanç and Zazaic is closely related to Goranî. The word **"Ankara"**, which means grape, is an expression that originates from Zazaic. In Zazaic, "(h)engure" means grape, so the official name of Ankara has been changed over time: **(H)Engure** (Zazaic, 700 BC) → **Angora** (Galatian language, 300 BC; Galatia's capital 278-25 BC) → **Anküra** (Greek, 200 AD) → **Ankara** (Turkish, 1200 AD).

My HomeDNA results show a migration path that corresponds with the trips of my family with mothers and fathers. The map shows how my DNA moved over time to a point where the last significant genetic mix took place on the gene pool that created my ancestral lines. Sometime after 484 AD my ancestors moved back to Zazadiyar (Tunceli, Bingöl and Muş) from northern Iran and when they reached there, the following is what they would have experienced.

Ebu Müslim al-Khorasanî led a rebellion in Khorasan in 747 that resulted in the fall of the Umayyad dynasty, which was then replaced by the Abbasid Caliphate. He was a great leader who understood how to exploit popular dissatisfaction and to drive anti-Ayyadic propaganda. After the establishment of the new Caliphate in Kufa, Ebu Müslim al-Khorasanî received Khorasan as an enfeoffment, but was later deposed by the second Abbasid caliph Ebu Cafer Mansur and murdered. Ebu Müslim al-Khorasanî was one of the most important figures as he carried on his activity in Khorasan. It is not known whether he was of Turkish, Arabic, Zazaic or Persian origin. This led the Shiites to look at it with suspicion and refuse to accept it immediately. However, he guided the Alevis and quickly became a popular and respected person in Mesopotamia. In connection with his participation for Alevis, the invasion of the Kırmanç people may have begun from Khorasan to Anatolia.

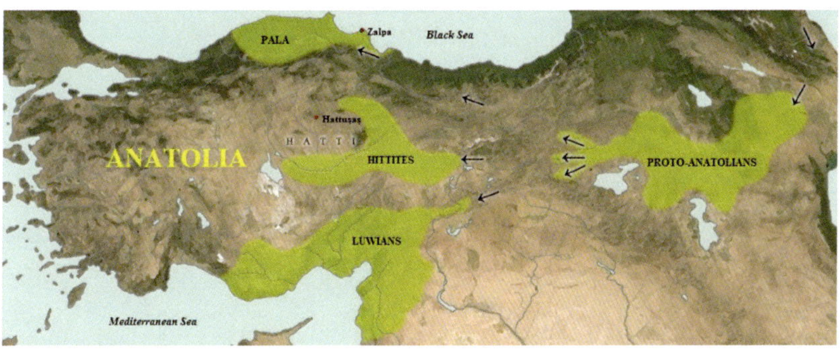

Figure 14. Proto-Anatolian Migrations 3000-2000 BC

In 965, Byzantine Emperor Nikephoros II conquered large parts of Eastern Anatolia and invited Armenian Christians to settle in the areas from which he expelled the Muslim population. Large groups of

Armenians obeyed the emperor's call, while virtually all Muslim Zazas in the provinces of Muş and Bitlis were driven to flee.

During the 11th century, the different Persian empires began to falter under the pressure of a new conqueror from the Turkmen steppes, the Seljuks. They took Khorasan, drove the Ghaznavids out of Iran and ended both the Buyid and Shia century of rule when they invaded Baghdad in 1055. Then, in Eastern Anatolia, the crucial battle of Malazgirt in the province of Muş occurred on August 26, 1071 between the Byzantine army under Emperor Romanos IV, which also included Armenians, Slavs, Goths and Germans, and the Seljuk army under Alp Arslan. The Seljuks won and Emperor Romanos was captured.

It was a crushing defeat for the Byzantines, which led to extensive religious and ethnic upheaval in Anatolia as the Muslim Seljuks gained control of the region. The battle led to the establishment of the Seljuk Sultanate in Anatolia and is considered the beginning of the end of the Byzantine Empire. For Byzantine, the loss of Anatolia was a fatal blow as the region was the most densely populated part of the empire. Unable to recover the lost territories, the new emperor Alexios I Komnenos was compelled to ask for help from Western Europe. Pope Urban II in Rome responded by urging Christians to liberate the Holy Land from Muslim rule, which led to the Crusades. According to an unproven claim, during this period the Koçgiri, Hormek, Hıran, Şadılı, İzol, Karsan, Arılı, Alanlı, and Demenan tribes have stayed in the Seljuks' wake from Khorasan and Nishapur in north-eastern Iran and settled in the valleys of the Upper Euphrates and Tigris.

As early as the 1510s, a new formidable enemy had appeared at Anatolia's eastern horizon – Yavuz Sultan Selim, who had become the ruler of the Ottomans. The prefix "Yavuz" is an expression of the cruel one. One of Yavuz Sultan Selim's first challenges as sultan was the growing tension with Iran's Shah Ismail I, who had recently taken the Safavids to power and switched the state religion from Sunni Islam to the "Twelver Imams" branch of Shia Islam. The Ottoman and Safavid armies met on the plain of Çaldıran (Van) in 1514. Although there were more foot soldiers in the Ottoman army, this was outweighed by the large number of cavalry on the Safavid side. However, the Safavid army had no cannons and the Ottomans had artillery forces. The battle of Çaldıran was historically significant because Shah Ismail I's reluctance to accept the advantages of modern firearms and the strategic importance of artillery was crucial. The battle ended in favour of the Ottomans. Shah Ismail I fled the battle-field when one of his soldiers gave him his horse. After such a heavy defeat, Shah Ismail I was unable to regain his previous reputation for invincibility, and so he posed no further danger to the Ottomans in Eastern Anatolia.

When the war broke out, it was more important than the Alevis who took a stand for Shah Ismail I. Revenge was completed when Zazadiyar fell in 1514 as a ripe fruit for Yavuz Sultan Selim's attacks. For decades, the Zaza people resisted the massive threat of the Ottoman colony. Then, the hunt for Alevis started. Yavuz Sultan Selim ordered the mass murder of **40,000** Alevis/Zazas, with the aim to end all Alevis/Zazas in the country. Subsequently, there was a mass deportation of Alevis/Bektaşis to different parts of Turkey and Balkan countries.

During the Turkish-Russian war of 1916-1918, the Zaza people were caught up in the crossfire. The villagers were defenceless. They had no weapons and they could not distinguish between enemies and allies. The provinces of Erzincan, Erzurum and Bingöl were badly affected by famine. The Russian and Armenian soldiers took all the food from the villagers. Domestic animals were easy prey and wheat was cut from the fields and roasted for the hungry warriors. Starvation led to a mass exodus of the population to the towns of the Black Sea.

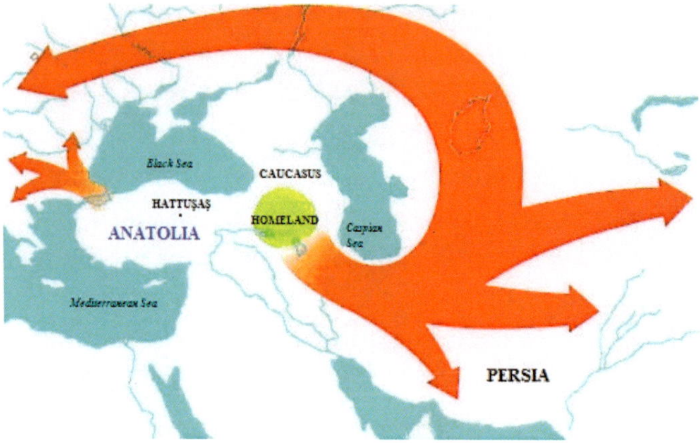

Figure 15. Indo-European Migrations in Eastern Anatolia 2000 BC

Dozens of Zaza tribes in the Dersim region of Eastern Anatolia rose in 1937–1938 against their oppressors. However, in the absence of mutual discipline and competent leadership, they succumbed to regular Turkish troops. Fighting against the well-equipped Turkish army is a long and difficult process, similar to eating soup with a knife. The Dersim rebellion was a Zazaic uprising against the Turkish government in the provinces of Tunceli, Bingöl, Elazığ and Muş. All of the tribes that participated in the rebellion were Zazas. The rebellion was led by Seyit Rıza, an Alevi Zaza chief in the Abbas Uşağı tribe. This region had a reputation for being rebellious, having been the scene of 11 separate periods of armed conflict over the previous 50 years. About 25,000 troops were deployed to stop the uprising. This task was completed during the summer and leaders of the uprising, including tribal leader Seyit Rıza and his 16-year-old son Resik Hüseyin, were hanged.

However, the remnants of the rebel forces continued to resist and the number of troops in the region doubled. The region was also bombed from the air. The rebels continued to resist until the region was pacified in October 1938. According to Osman Pamukoğlu, a general in the Turkish army in the 1990s, Kemal Atatürk had given the operative order himself. According to an official report of the Fourth General Inspectorate, **13,160** civilian Zazas were killed by the Turkish army and depopulated the province and about **14,400** Zaza people were forced to move to other parts of Turkey. The number of Zaza population exiled by province – Konya 1264, Balıkesir 1087, Manisa 1015, Aydın 1003, Bilecik 866, Bursa 861, Çankırı 750, Denizli 659, Eskişehir 643, Kütahya 586, İzmir 511, Antalya 410, Amasya 392, Kastamonu 386, Afyon 356, Çanakkale 321, Uşak 273, Yozgat 251, Kayseri 245, Bolu 223, Isparta 181, Çorum 171, Muğla 170, Nigde 124, Sinop 110, Tekirdağ 100, Kırklareli 60, Burdur 54, Edirne 43, İstanbul 13 and Zonguldak 7 are persons. It caused the most violent torture ever seen in an uprising during the Republican years. Those who did not participate in the uprising and the families of the rebels were also tortured.

In 1961, Western Europe opened its doors to poor people in Turkey to be employed in various jobs as labourers. Poor small farmers and workers to be employed in heavy work (in mining, construction, various manufacturing sectors) from Eastern Anatolia began to migrate to the gold country ***El Dorado*** in the West.

Almost all of the workers migrating from Turkey were masses with no professional experience, unskilled and with a very low literacy rate. In addition to Turks, the majority of these masses were Zaza and Kurdish peoples. Guest workers migrating from Turkey to Western Countries were literally forced to work as slaves in heavy work in the Fourth Empire.

The migration from Turkey to the West created a domino effect. After Adolf Hitler's (the famous superior race) Third Empire, the cry he uttered in the Fourth Empire (Industrial Empire) was based on labour. In other words, the logic of doing more work for less money was dominant. The countries in need of cheap labour were West

Germany, Austria, Switzerland, the Netherlands, Belgium, France, Denmark and Sweden and so on.

The working conditions of workers migrating from Turkey in heavy industry were regulated without even the smallest human rights being disregarded and without the need to create some positive conditions, even if at a minimal scale, and to protect them by law. The work these people did was harmful to health, dirty, stressful and tiring. Since everything in foreign countries was pork, the food for expats was uniform, such as bread, potatoes, chicken, olives, beans and butter.

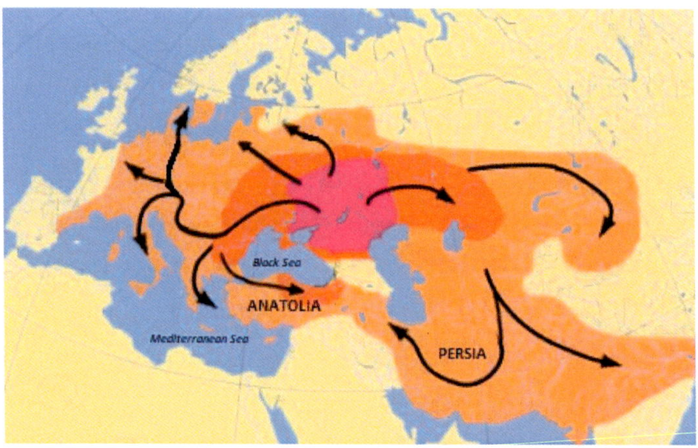

Figure 16. Indo-European Migrations 4000-1000 BC according to the Kurgan Hypothesis

These workers had never encountered any agricultural pesticides in food before, and were not familiar with products produced with artificial fertilizers in their own countries, and were also not exposed to environmental pollution and exhaust gases in their own countries. Due to their diet, environmental conditions and most importantly the work they did, they began to suffer from cancer, as well as cardio-vascular diseases. They could not tolerate food chemicals and their immune defence system was very weak compared to city dwellers. To tell the truth, a city dweller has been exposed to harmful chemicals in food since childhood. Doctors would prescribe prescriptions for these workers without examining them. The cupboards full of medicines in their homes did not treat the workers. Most of the workers who migrated from Turkey died before reaching the age of 70.

When Turkish expatriates returned to Turkey, they realized the thoughts and dreams they had worked hard for and put in a lot of effort for the future. They saw that they had put in a lot of effort to create financial security for themselves and their relatives. It was a situation of sacrificing a lifetime for their dreams of becoming rich. Work, suffer, earn and save. This was their motto. It did not take many years for these poor workers from Anatolia to reach the level of being very rich. While the investments in the country were mostly concentrated in real estate, those who had not previously owned donkeys in their own country now had luxury cars such as Mercedes, BMW, Audi and Porsche. Special importance was given to the fact that every visit to the homeland was definitely made in a five-star hotel.

Their wealth took them to a more distant world. With the pride of their own people, everyone claimed to be the genius of nature. The behaviour of the guest workers went beyond what was appropriate or allowed. In desperate attempts, the one who knew who was the most and the alpha male in the hierarchy of dominance would be. Their arrogant behaviour often led to a conflict between people in the atmosphere of hatred, contempt and hostility. Based on some political purposes and because of the thought, religion and sect conflicts that created the reason, a conflict created between the children of the same race, the guest workers who came from the same origin and had common interests continued for years. People all over Turkey encountered the arrogance symptoms of Germans. The unique feature of the person consists of the issue of wealth/poverty, competition and arrogance.

THE DNA OF THE ZAZA PEOPLE

Turkey encompasses a variety of ethnic groups and minority groups. However, there is very little official information on many of its main minorities. The Zazas comprise the third largest ethnic group in Turkey. Tolerance against the ethnic minorities Zazas and Kurds has always been low in Turkey. The Zaza people have often considered themselves victims in the Ottoman Empire. Their own literary works have never existed because the Zazaic language lacked an alphabet and all forms of print were forbidden. However, the oral tradition and narrative art are significant and often presented in the form of a storyteller.

For a long time, one could not even mention publicly that there was an ethnic group in Turkey called Zazas – instead along with the Kurds, they were called **"Mountain Turks"**. However, in the 1960s, Turkey's president Cemal Gürsel shocked the establishment by speaking openly about Zazas and Kurds. Turkey's fourth president Cemal Gürsel was born in Hınıs in Zazadiyar. His mother was Zaza and his father was Alevi Turk. The four-star general and commander-in-chief had previously kept his ethnic origins secret in order to maintain his military career. During Cemal Gürsel's presidency, this secrecy eased, which led to greater attention on the ethnic groups. For the first time, one could openly identify as Zaza or Kurd. A gendarmerie group was appointed to keep statistics on Zaza inhabitants in Eastern Anatolia. As noted earlier, the gendarmerie discovered that by 1970, a total of 350,000 Zaza inhabitants lived in Eastern Anatolia. These statistics are rather striking as there were many tribes and places missing. More realistically, the figure would have been **500,000**.

The starting shot came after Cemal Gürsel's presidential entry. It became crowded in authorship, writer after writer received a free ride. From all corners of Anatolia, Zaza writers frequently wrote in their works that the Zaza people are **Turkic people**. They wrote a complete series of unpolished truths. With the philosophy of ethnic origin, the authors built bridges between fire and water. Nothing is known about the details of the facts as there was no tangible evidence and no one can prove that this is right. They made little of themselves. These authors fabricated that all Turkic people who existed are ancestors of the Zaza tribes Türkmen, Oğuz, Avşar, Yörük, Gagavuz, Harzem, Hun, Kıpçak, Varsak, Türgiş, Kınık, Yıva, Eymür, Yazır, Döğer, Beydili, Bayındır, Kaçar, Bozulus, Kayı, Cerid, Bayad, Kızık, Çiğil, Karluk and İğdir. Out of nowhere, a Zaza tribe was assigned to four different Turkic people, for example, the **Alanlı** tribe of Oğuz, Avşar, Eymür and Yıva people. The basis for the supposition is primarily to make Zaza Turkish. Many writers claim that Zazaic is a Turkish language but are not able to provide evidence. Attempts have been made to prove the relationship of the Zaza language with Turkish without convincing success.

Against all odds, I actually learned about my provenance after trying six DNA tests. Results can vary widely depending on which test company one uses. I have been interested in my origins for a long time, not only to understand my genetic ancestry but also to see how closely all Zazas are linked to each other. I knew in principle that our roots were in **Western Iran** and **Eastern Anatolia**. I also knew that a company can only match people to relatives in their customer base, so if I wanted to find as many relatives as possible, I would have to use several companies. Therefore, I sent my DNA samples to Ancestry-DNA, 23andMeDNA, MyHeritageDNA, Family TreeDNA, Living-DNA and HomeDNA. All the DNA samples were intended for autosomal testing and test companies 23andMeDNA and LivingDNA included Paternal Haplogroup Y-DNA and Maternal Haplogroup mtDNA analyses.

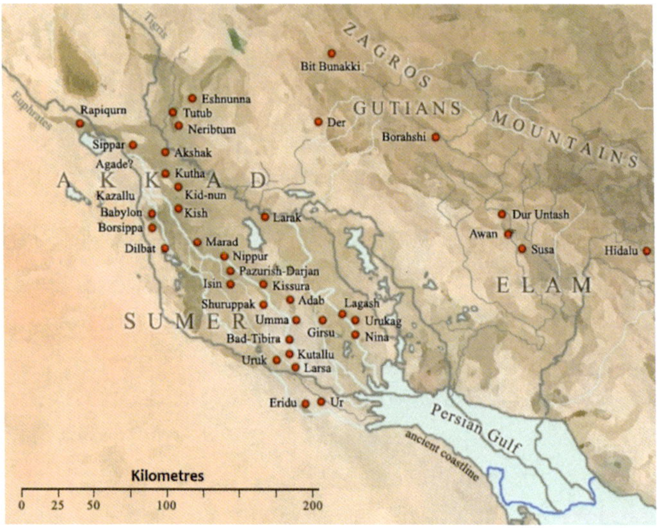

Figure 17. Map of Elam and Sumer

Explanation of DNA

DNA is an abbreviation that stands for the chemical name "Deoxyribo Nucleic Acid". It is the chemical substance that carries the genetic code, a code that contains the recipe for all life forms. The DNA code is basically simple – it only consists of four building blocks, which are abbreviated by the letters **A**, **C**, **G** and **T**. The code itself, on the other hand, is very long. In total, there are about three billion nucleotides (letters) in our genome. The strand of nucleotides is almost identical between individuals, but on average a nucleotide of a thousand differs in a comparison between two randomly selected individuals. The deeper DNA test reveals the genealogical traces stored in our Y-Chromosome. Traces left behind by our ancestral line for thousands of years. These traces are then compared with the DNA of other people tested. Based on similarities and differences, it is possible to draw a tree of what the relationships between the people look like. A strange family tree emerges that spans 2,500 years. A family tree, that originates from a joint that lived during the Iron Age. About 95% of our DNA is so-called autosomal DNA, whether you are male or female. It is this DNA that we randomly inherit from both mother and father. We have thus inherited the autosomal DNA we carry from both our parents and the millions of puzzle pieces can come from any branch of our pedigree. The puzzle pieces of DNA that we find in the autosomal DNA are called segments. The length of a segment (chromosome) is measured in a special unit called centiMorgan and abbreviated cM.

Origin and DNA Tests

My user experience and results were quite different for each company I sent my samples to. The results are generic and the ethnicity categories are too broad. The companies offer detailed ethnicity estimates for people of origin from the **British Isles**, **Germanic Europe**, **Fennoscandia**, the **Iberian Peninsula**, **South Europe**, **East Europe** and **Russia**. These ethnicity estimates are created for Americans of European origin who are interested in finding out where they came from, knowing they are immigrants to America. The test companies did not appear to have entered deeply into the material for **Middle Eastern** origins. They lacked information in their customer base for most existing ethnic groups in the Middle East. Eastern Anatolia is archived in world history with the attribute "Armenia". Zaza candidates often feel offended when DNA results emerge as "Armenian descent", "Jewish descent" and "Greek descent".

The ethnicity map in the test results seeks to highlight an old-fashioned and outdated view of ethnicity, ancestry and racial affiliation. According to Europeans, ethnicity maps have been added to satisfy an American target group. One of the companies that performs DNA testing with ethnicity maps notes that one should not compare the tests with a thorough science. It is believed that the DNA tests are useful when people are looking for their relatives or want to answer a family riddle but the ethnicity map does not reveal much.

My DNA results from HomeDNA showed that my ancestors came from south-eastern India, western Siberia, Fennoscandia, Sardinia and so on – which is highly unlikely. Therefore, these results are not included in the DNA statistics. Test companies LivingDNA and HomeDNA did not offer any matches for relatives. They do not currently provide any hit lists, but one only knows it on one's own behalf – one cannot compare oneself with anyone. Most test companies provide no information about ancient ancestry and do not adequately explain the results.

More and more Zazas allow DNA testing to investigate their family background and ethnic origin. So far, most Zaza candidates have been tested on MyHeritageDNA (Western Europe) and 23andMeDNA (USA). The majority of Zaza candidates live in Western Europe, the United States and Turkey. The age range is 20-40 years and covers about 80% of candidates. AncestryDNA and 23andMeDNA tests on Middle Eastern candidates are, in many ways, more affordable to work with.

Taking the initiative to delve into the way of the Zaza people through history is a staggering thought. The Zazas and the Alevis in Turkey lived in a tribal society and applied strict **endogamy**. In fact, we Zazas are intermarried and it is enough to go back only 2,500 years in time for the Zaza people to find a common ancestor. The key to finding a relationship between DNA matches is to compare our home regions, the heartland. The phenomenon is common in regions where the genera are particularly intermarried and closely related to each other, such as in the provinces of Tunceli, Erzincan, Bingöl and Muş.

Many tantalizing challenges and interesting discoveries await as one enters the exciting world of DNA genealogy. DNA tests can help discover fascinating family histories and solve the most difficult genealogy mysteries. Using the tool "Are your parents related?" on GEDmatch, I did a test that indicates my parents are genetically very close to each other. My parents share common DNA 91 centiMorgans, which indicate they are "third cousins". This was a startling discovery, as I had no idea that my parents would be related to each other.

The Matches

DNA tests show with fascinating clarity the kinship that ties the Zaza tribes together. More than **11,500** potential DNA relatives emerged in my results. DNA tests show a calculation of how close Zazas are. The challenge for me was to find out exactly how we are related. The most frequent provinces in Eastern Anatolia for DNA matching are Tunceli, Erzincan, Bingöl, Sivas (İmranlı, Zara, Kangal), Muş (Varto), Erzurum, Elazığ, Gümüşhane (Kelkit) and Bayburt. The small town of Pülümür in Tunceli is over-represented on DNA tests, with tribes such as Lolan, Balaban, Arılı, Baba Mansur, Çarekli, Kemanlı, Rotanlı, Sisanlı, Şavalanlı and Şeyh Hasanlar. Generally speaking, all **220** Zaza tribes have participated in DNA testing. The tribes most represented in tests are Koçgiri, Hormek, Lolan, Balaban, Şadılı, Arılı, Alanlı, Haydaran, İzol, Baba Mansur, Kureyş, Derviş Cemal, Şeyh Hasanlar, Abbas Uşağı and so on.

My common matches with Zaza cousins were taken into account at thresholds between 15-1,800 centiMorgans. Matches with distant

DNA cousins represented thresholds between 5-20 centiMorgans. My distant DNA cousins from all over the world included Kurds, Turks, Lazes, Armenians, Greeks, Iranians, North Americans, Eastern Europeans, Circassians, Arabs, Georgians, Afghans, Jews and so on. A trend among the Zazas who became Sunni Muslims after 1514 showed low shared DNA matches of 20-40 centiMorgans. This may be due to mixed marriages with other ethnic Sunni Muslims, such as with Kurds, Turks and Lazes. Matches with my Khorasan cousins were between 15-30 centiMorgans. The emigrant Goranî cousins more than 2,800 years ago were with my common matches between 15-25 centiMorgans.

In the 1510s, Zazas/Alevis were deported to the provinces of Tokat, Amasya, Yozgat and Çorum, and they align with my matches ranging 30-40 centiMorgans. The Zaza tribe Ömeran was deported from the provinces of Tunceli and Bingöl to the provinces of Diyarbakır and Mardin. In 1850, they then migrated to Konya district and account for my shared DNA matches ranging 25-40 centiMorgans. I belong to the Hormek tribe. Mixed marriage from the 15th century between Hormek and other Zaza tribes with my shared matches is between 80-240 centiMorgans. The Hormek tribe has a high proportion of mixed marriages with tribes such as the Balaban, Lolan, Arılı, Maskan, Koçgiri, Şadılı, Menikan, Demenan, Karsan, Şıh Hasan, Adaklı, İzol, Bahtiyar, Arey, Haydaran, Laçin Uşağı, Çarekli, Hıran, Sarı Saltık, Kureyş, Alanlı, Şavalanlı, Keman, Karaballı, İbikiyan, Abdalan, Abasan, Baba Mansur, Derviş Cemal, Ağu İçen and so on. The majority of these tribes belong to the provinces of Tunceli, Bingöl, Erzincan, Muş and Sivas. My Armenian cousins across the globe have their roots in Varto, Kığı, Palu, Hozat, Hınıs, Çemişgezek, Kemah, Elazığ, Erzurum, Sivas, Bitlis and Van.

The Zazas who entered into a mixed marriage with North Americans and Western Europeans (mining and automotive workers in the United States) 100 years ago have a quarter of their Zaza ethnicity remaining in the genome, showing just over 40 centiMorgans in my DNA match. From the North Caucasus came the Circassians, who settled in Varto (Muş) in 1864 and have entered into mixed marriages with the Zaza people showing 15 centiMorgans. Many Armenians are half Zaza (Bingöl, Elazığ, Muş, Tunceli), but unfortunately, they categorically deny their origin. The test company AncestryDNA clearly establishes Iranian/Persian ancestry for Armenians of 0-5% and Zazas as 45-75%.

Using methods for identifying common ancestors "Advanced DNA Comparison", "Chromosome Browser", "One-To-Many DNA Comparison Result" and "Triangulation", it could be stated that intermarriage between the "Evladî-Resul" strains is very high. Virtually all of these "Dede" clans are very closely related to each other and have common ancestors due to intermarriage and their mutual descent. It seems that the "Evladî-Resul" tribes were well thought out and avoided marrying the other Zaza tribes according to the incest taboo. In particular, it is strictly forbidden to marry between the "Dede/Seyit" families and the "Talip/Mürit" families.

The result that I also had many relatives in the Balkans puzzled me at first, as I did not know that I would have any Eastern European roots. I dug into this. Did my own family come from the Balkan countries, or was it the other way around? Research and proven experience that Janissaries (élite troops of the Ottoman Empire) and Bektaşis (the Alevi Sufi Dervish Order) were refugees in 1514 accounted for the traces in the DNA I carry. It was the Alevis who founded the Janissary army corps in 1361. Within the Janissary army, many of the Alevis held key command posts such as "Yeniçeri Çavuşu" (Janissary sergeant). Thanks to the Janissary troops, the Ottoman Empire expanded its borders in Europe and won war after war. Thus, it is possible that many of the Janissary soldiers of Zaza descent did not return to Anatolia after the constant wars in Europe, such as the Battle of Vienna (Battle of Kahlenberg) in 1683.

Ethnicity Estimates

The Surname Law was introduced in Turkey in 1934. The reason for this is that no residents of Turkey have a patronymic or metronymic. Otherwise, it could have been interesting to be able to quickly check whether we Zazas have any common family names where we could start searching for relationships. To determine my relatives, my DNA profile has been compared to about **3,500** Zaza individuals from around the world. The compilation tables below present the average values of the estimated ethnicity distribution of the Zaza candidates for different DNA test companies. DNA results of mixed marriages with the Zaza people and the very distant cousins that shared DNA matches less than 15 centiMorgans were not included in the statistics. The test company MyHeritage's categories of ethnicity are too broad and destroy the facts with a half-complete and highly divergent result. The test company AncestryDNA seems to be an optimum tool for the ethnic categories of the Middle East.

Figure 18. Map of Eastern Anatolia

The following table presents the average values of 3,500 Zaza DNA candidates from around the world:

Ethnicity Estimate	**Ancestry**	**23andMe**	**MyHeritage**	**FamilyTree**	**LivingDNA**
Iran/Persia	**59.4%**				
Turkey & the Caucasus	**39.8%**				
Middle East	0.4%		3.2%	**88.2%**	
Southern Asia	< 0.1%	0.1%	2.2%		
Central & Northern Asia	0.1%				
West Asia		**98.4%**	**86.1%**		
Europe		0.8%			
North Africa			1.7%		
East Asia	< 0.1%	0.4%			
North and West Europe			1.2%		
Sub-Saharan Africa		0.1%			
South Europe	0.1%		3.9%		
Central Asia			1.1%		
East Europe			0.6%		
Central & South Asia				**8.4%**	**10.2%**
Southeast Europe				2.4%	
Near East					**89.8%**
Unassigned		0.2%		1.0%	

The following table presents my own DNA results from different test companies:

Ethnicity Estimate	**Ancestry**	**23andMe**	**MyHeritage**	**FamilyTree**	**LivingDNA**
Iran/Persia	**52.0%**				
Turkey & the Caucasus	**44.0%**				
Middle East	2.0%			**90.0%**	
Southern Asia	1.0%				
Central & Northern Asia	1.0%				
West Asia		**98.3%**	**82.0%**		
East Asia		1.2%			
Europe		0.1%			
South Europe			9.6%		
Central Asia			6.3%		
East Europe			2.1%		
Central & South Asia				**7.0%**	**11.2%**
Southeast Europe				3.0%	
Near East					**88.8%**
Unassigned		0.4%			

The compilation of autosomal DNA of 3,500 Zaza candidates has been performed as follows:

- **West Asian 93.1%** = West Asia (98.4% + 86.1%) + Iran/Persia (59.4%) + Turkey & the Caucasus (39.8%) + Middle East (88.2% + 3.2% + 0.4%) + Near East (89.8%).
- **Central & South Asian 4.4%** = Central/South Asia (10.2% + 8.4%) + Southern Asia (2.2% + 0.1% + 0.1%) + Central Asia (1.1%) + Central & Northern Asia (0.1%).
- **European 1.8%** = South Europe (3.9% + 0.1%) + Southeast Europe (2.4%) + North and West Europe (1.2%) + Europe (0.8%) + East Europe (0.6%).
- **North African 0.4%** = North Africa (1.7%) + Sub-Saharan Africa (0.1%).
- **East Asian 0.1%** = East Asia (0.4% + 0.1%).

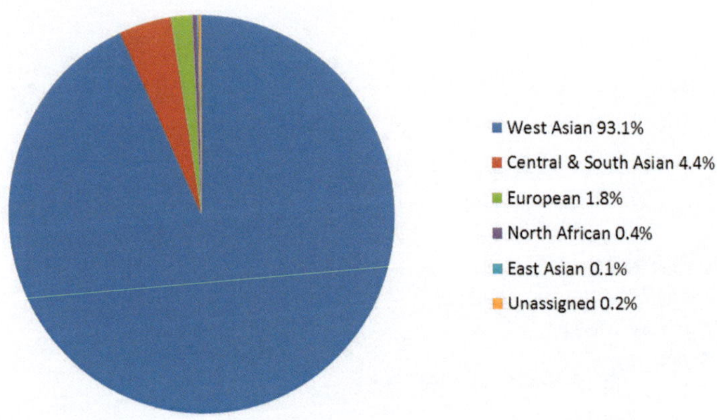

Figure 19. Average of Autosomal DNA of 3,500 Zaza Candidates

The percentages represent the proportion of MyHeritageDNA users in Turkey who have this ethnicity: West Asian 31.8%, Turkish (Central Asian) 12.5%, Middle Eastern 11.1%, Italian 9.0%, Balkan 8.8%, North African 5.5%, Eastern European 5.3%, Greek 5.1%, Western European 2.3%, Scandinavian 2.0%, Baltic 2.0%, English

1.6%, Iberian 1.4%, South Asian 0.8% and Irish 0.8%. In Turkey, there are about 42 different ethnic groups. However, all the calculations of their sizes are disputed. Additionally, there are political and sometimes ethical problems in attempting to categorize the population into ethnicities. The largest ethnic groups are Turks, Kurds, Zazas, Lazes, Georgians, Arabs, Circassians, Bosnians, Gypsies, Albanians, Jews and so on.

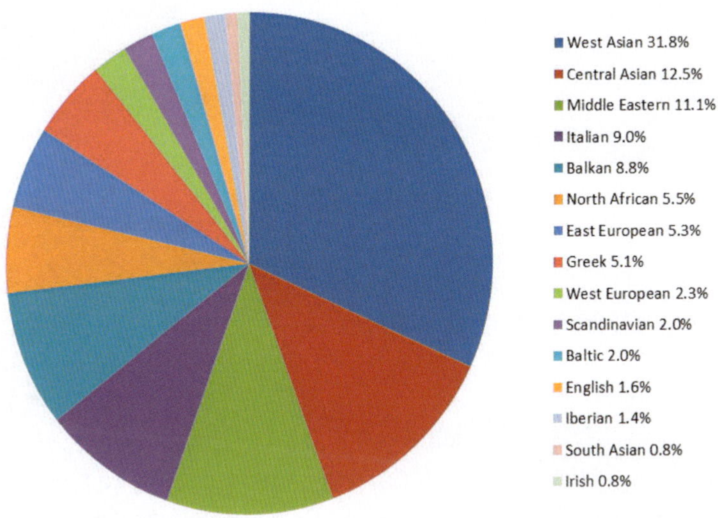

Figure 20. Autosomal DNA of Population in Turkey according to MyHeritageDNA

My ancient DNA matches can be compared to populations around the world with the tool “Archaic DNA Matches” on GEDmatch. The upper limit threshold was adjusted to 1 centiMorgan. The ancient specimens range in age from approximately 1,100 to 50,000-year-old finds from archaeological sites in Europe, Asia (Siberia) and America. Each country represents a number of matched finding places. These skeletal discoveries appear to have their roots in Anatolia. The **Denisova** discovery in Siberia, of ancient human remain that are 30,000 years old, indicates a very short DNA bit on segment 22 (Family Tree and MyHeritageDNA).

Country	Age	Ancestry	23andMe	MyHeritage	FamilyTree	LivingDNA
Russia	2000-50000	21	14	21	22	22
Germany	1100-7000	25	12	25	21	22
Hungary	2900-7200	9	3	9	9	9
Sweden	1500-7000	6	3	6	6	3
UK	1300-2000	2	1	4	4	3
USA	8300-12600	2	1	2	2	1
Czech Rep	~ 2000	2	1	2	2	2
Spain	~ 7000	1	1	1	1	1
Luxemburg	~ 8000	1	1	1	1	1
Poland	~ 3400	1	1	1	1	1
Estonia	~ 2000	1	1	1	1	1
Greenland	~ 4000	1	1	1	1	1

The Haplogroups

I carry 23andMe Paternal Haplogroup Y-DNA **R-L23** and Maternal Haplogroup mtDNA **J1,** as well as LivingDNA Haplogroup Y-DNA **R-M269 (R1b)** and Maternal Haplogroup mtDNA **J1 (J1d5a)**. The Y-DNA R-M269 (R1b) itself is strongly associated with the Neolithic extensions. When Anatolian farmers brought farming to the European continent, they often moved and bred native hunter-gathering populations. It is believed to have existed somewhere between **4,500** and **9,000** years ago and is therefore a highly useful Haplogroup for geneticists and archaeologists who wish to trace the migration patterns of these agricultural populations. Today, an estimated 110 million European men carry the Y-DNA R-M269 father, with the highest frequencies found in the West. This follows the spread of agriculture, which would have entered Europe from Anatolia before spreading into France and the United Kingdom. The frequency of Y-DNA R-M269 decreases towards Eastern Europe as the frequency of Y-DNA R1a fader lines increases. Only 10-15% of men in Turkey carry Y-DNA R-M269 and about 9% of Zazas worldwide carry Y-DNA R-M269.

For men with paternal Zaza origins, it is most common to belong to Haplogroup **R** (about 51%), followed by **J** (about 27%), **E** (about 8%), **G** (about 4%), **L** (about 3%), **Q** (about 2%) and **C**, **H**, **I**, **N**, **T** (each about 1%).

For both women and men with maternal Zaza origins, it is most common to belong to Haplogroup **H** (about 34%), followed by **U** (about 25%), **J** (about 14%), **T** (about 10%), **R** (about 5%), **K** (about 4%), **B**, **D**, **I** (each about 2%) and **N**, **X** (each about 1%).

Many ancient human skeletal discoveries have been found in the regions around Zazadiyar. In Anatolia, the DNA of six of these ancient remains has been examined. The results of the DNA study belonging to the Anatolian prehistoric peoples show the following:

- Find 1: Karaman "Epipaleolithic" 13642-13073 BC Paternal Haplogroup Y-DNA **C1a2** and Maternal Haplogroup mtDNA **K2b**.
- Find 2: Çatalhöyük "Pre-Pottery Neolithic" 8300-7800 BC Paternal Haplogroup Y-DNA **G2a2b2b**, **C** and Maternal Haplogroup mtDNA **U3**, **K1a**, **N1a1a1**.
- Find 3: Niğde "Pottery Neolithic" 7500-5800 BC Maternal Haplogroup mtDNA **K1a12a**, **N1a1a1**, **N1b1a**.
- Find 4: Bursa "Pottery Neolithic" 6400-6000 BC Paternal Haplogroup Y-DNA **J2a**, **H2**, **G2a2b2a** and Maternal Haplogroup mtDNA **K1a4**, **N1a1a1**, **N1b1a**, **J1c11**.
- Find 5: Bursa "Pottery Neolithic" 6400-5600 BC Paternal Haplogroup Y-DNA **G2a2a**, **I**, **G** and Maternal Haplogroup mtDNA **X2m2**, **K1a4**, **N1a1a1**, **H**, **K1a2**.
- Find 6: Malatya "Late Chalcolithic" 3941-3031 BC Paternal Haplogroup Y-DNA **J2a1a**, **H2** and Maternal Haplogroup mtDNA **J1c16**, **T1b**.

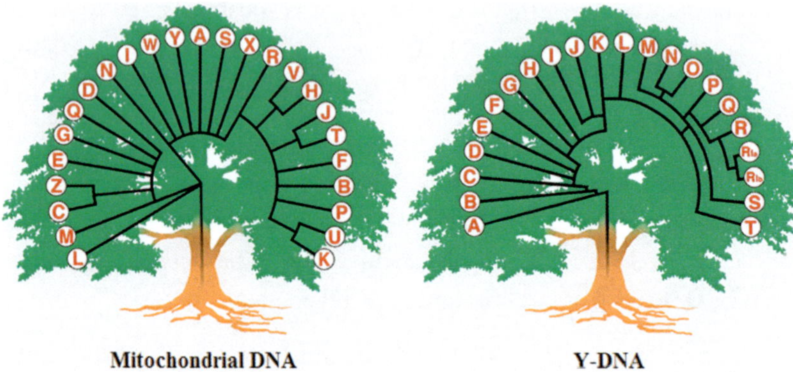

Figure 21. Haplogroup mtDNA Tree and Haplogroup Y-DNA Tree

Four ancient skeletal discoveries have been found in the Kermanshah and Mahabad districts of Western Iran. The results of the DNA study belonging to the prehistoric peoples reveal the following:

- Find 1: Kermanshah "Early Neolithic" 8200-7700 BC Maternal Haplogroup mtDNA **R2**, **T2c**.
- Find 2: Kermanshah "Early Neolithic" 7500-7000 BC Paternal Haplogroup Y-DNA **G2b** and Maternal Haplogroup mtDNA **J1d6**.
- Find 3: Kermanshah "Neolithic" 5900-5600 BC Paternal Haplogroup Y-DNA **G2a1** (xG2a1a) and Maternal Haplogroup mtDNA **K1a12a**.
- Find 4: Mahabad "Iron Age" 971-832 BC Paternal Haplogroup Y-DNA **R1b1a2a2** and Maternal Haplogroup mtDNA **N1a3a**.

In large parts of the Caucasus, DNA analyses of nine ancient skeletons have been performed. The results of the DNA study belonging to the Caucasian prehistoric people reveal the following:

- Find 1: Kutaisi "Satsurblia" 11400-11100 BC Paternal Haplogroup Y-DNA **J** and Maternal Haplogroup mtDNA **K3**.

- Find 2: Kutaisi "Kotias" 7900-7500 BC Paternal Haplogroup Y-DNA **J2a** and Maternal Haplogroup mtDNA **H13c**.
- Find 3: Pyatigorsk "Eneolithic Steppe" 4336-4047 BC Paternal Haplogroup Y-DNA **R1a1a** and Maternal Haplogroup mtDNA **I3a**, **H2**.
- Find 4: Nakhchivan "Chalcolithic" 4300-3700 BC Paternal Haplogroup Y-DNA **L1a** and Maternal Haplogroup mtDNA **H**, **H2a1**, **K1a8**.
- Find 5: Vladikavkaz "Maykop" 3695-3545 BC Maternal Haplogroup mtDNA **J2a1**.
- Find 6: Gyumri "Kura-Araxes" 3631-3030 BC Paternal Haplogroup Y-DNA **G2b** and Maternal Haplogroup mtDNA **R1a1**, **K3**.
- Find 7: Ashtarak "Kura-Araxes" 3400-3100 BC Maternal Haplogroup mtDNA **U3a2**.
- Find 8: Pyatigorsk "North Caucasus" 2863-2581 BC Paternal Haplogroup Y-DNA **R1ba2** and Maternal Haplogroup mtDNA **H13a1a2**.
- Find 9: Chambarak "Kura-Araxes" 2600-2400 BC Paternal Haplogroup Y-DNA **R1b1-M415** and Maternal Haplogroup mtDNA **H1u**, **X2f**.

The results of the DNA study of Haplogroup mtDNA **R2**, **J1d6** and Haplogroup Y-DNA **R1b1a2a2** from Kermanshah and Mahabad in Western Iran can be matched with Zaza people's Paternal origin Haplogroup **R** (about 51%) and **J** (about 27%) and Maternal origin Haplogroup **J** (about 14%) and **R** (about 5%). The Zaza people's matches belong to Haplogroup **R1b** (R-M269) and Haplogroup **J1**, which proves that Zazas count their kinship around 6,500 years back in time. However, the results of DNA studies that treat prehistoric Anatolian and Caucasian peoples do not adequately follow the template of kinship with Zaza people's DNA tree. On the other hand, the skeletal finds in Eastern Anatolia in Malatya and Urfa with

Paternal Haplogroup **J2a**, **J2b** and Maternal Haplogroup **U1a**, **J1c** agree to some extent consistent with the DNA trees of the Zaza people.

The Scythians appear for the first time in historical sources in the late 800s BC when they migrated to Eastern Anatolia. The Scythians left thousands of kurgans (burial mounds). Many ornaments found in these burial mounds depict Scythian everyday life. The genetic variation within the Scythian nomadic group is so broad that it must be noted that they attracted people they came into contact with. The central Eurasian steppe where the Scythians came from seems to have been a highly dynamic place. This important geographical region served as both a melting pot and nursery for people as well as for communicative and technological innovations. Here people met and shared ideas and genes. These ancient people belonged to Paternal Haplogroup Y-DNA R-M17, R1a and Maternal Haplogroup mtDNA N1, F2a. The Zaza people belong to about 4% of Paternal Haplogroup Y-DNA R-M17.

The DNA test company 23andMe presents Neanderthal heritage in terms of the number of genetic variants carried. I have more 253 Neanderthal variants than 32% of 23andMe customers. However, my Neanderthal ancestors account for less than 4% of my total DNA. Close cousins (about 100 people) have, on average, more 234 Neanderthal variants than 11% of 23andMe customers.

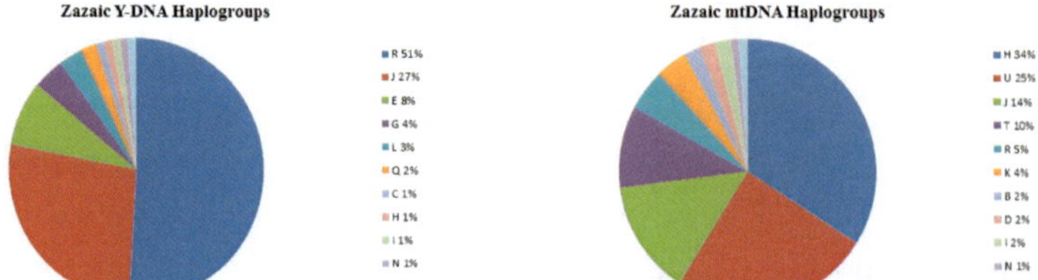

Figure 22. Zazaic Y-DNA and mtDNA Haplogroups

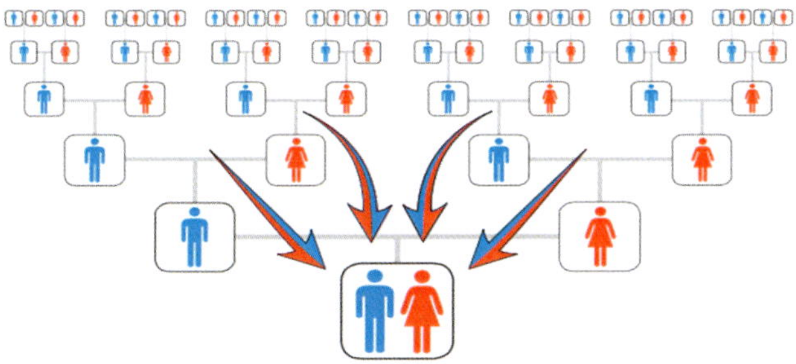

Figure 23. Autosomal DNA Archives

Test Results

Eastern Anatolia is dominated by the overlap of ancient high cultures. Aboriginal inhabitants were Sumerians (4000 BC) and Hittites (3000 BC). It was the scene of many invasions of different empires that replaced each other. The ancient Iranian ethnic groups in the region were the Elamites (3200 BC), Gutians (3000 BC) and Parthians (1000 BC). From the Caucasus came the Hurrians (3000 BC), Medes (2000 BC), Cimmerians (1400 BC), Urartians (1200 BC) and Scythians (800 BC).

The test company 23andMe found the strongest evidence of my provenance in the following ten provinces in Turkey: 1 Tunceli (strongest), 2 Erzincan, 3 Sivas, 4 Elazığ, 5 Bingöl, 6 Muş, 7 İstanbul, 8 Malatya, 9 Kayseri and 10 Osmaniye (weakest). In addition, there is a province of Razavi in north-eastern Khorasan in Iran that provided probably proof of ethnicity. Major migrations of the Kırmanç people eastwards to Khorasan in Iran appear to have taken place in 800 BC. As noted earlier, Northern Zazas are referred to as Kırmanç (Kurmanç). Zaza people's paternal land includes the provinces of Tunceli, Bingöl and Muş. Erzincan was populated by Zazas in the 13th century during the epoch of the Seljuks and an exodus to Sivas took place in the 1510s after massacres of Alevis/Zazas by Yavuz Sultan Selim. After mass executions of Zazas in Dersim in 1938, eight Zaza tribes were forced to move to Konya, Balıkesir, Aydın, Bilecik, Kayseri, Yozgat, Amasya and Çorum.

The test results of the five companies prove ethnicity for Zazas from around the world in **West Asia**, **Iran/Persia**, **Turkey & the Caucasus**, **Middle East** and **Near East**. Eastern Anatolia was the land of the ancient Elamites, Hittites, Sumerians, Hurrians, Urartians and so on. The migration of the Mesopotamian ethnic groups of the Elamites and the Sumerians to north-western India (Rajasthan) around 2200 BC provides the basis for the **Central/South Asian** ethnic origin. Iran's old inhabitants included people and cultures that were among the world's earliest farmers and built "kerpiç" adobe houses.

A well-known civilization on Iran's western borders was **Elam**, whose people built cities, had a written language and fought with ancient civilizations. The history of the land of Elam can be traced back to the fourth millennium BC. The region was then covered by the Mesopotamian Uruk culture. In about 3200 BC, the proto-Elamite period began. From that time come texts that are among the absolute oldest known in the world. This proto-Elamite writing bears a strong resemblance to the Sumerian's first cuneiform. It has been assumed that the script spread from Sumer to Elam, but it may just as well have been the reverse. A cultural boundary passed between the Elamites who inhabited the floodplain in the lowlands and those who lived up in the more remote highlands.

Lowland culture was more strongly influenced by contemporary Mesopotamian cultures in the West. The life of the agricultural community in the lowlands was more like the life in Mesopotamia compared to the different climatic conditions of the highlands. Elam had rich natural resources in the form of metals, stones and timber. These were particularly sought after by the people of Mesopotamia who lacked these natural resources. The two neighbouring countries seem to have favoured conquest and looting as a way to obtain each other's valuable natural resources. The Sumerian civilization dates back to around 5000 BC.

The first Sumerians were farmers who worked hard to gain control of the Euphrates and Tigris rivers. They wanted to be able to use the rivers' water to give life to the crops and also try to prevent flooding. The Sumerians needed more land to cultivate, so they dug trenches in fertile Mesopotamia. Thus, wherever there was abundance, they gained more and more cultivated land. As the Sumerians cultivated wheat, they made canals or ditches, which they filled with water from the nearby Euphrates and Tigris rivers. These efforts seem to have created a larger and more organized civilization than other agricultural societies.

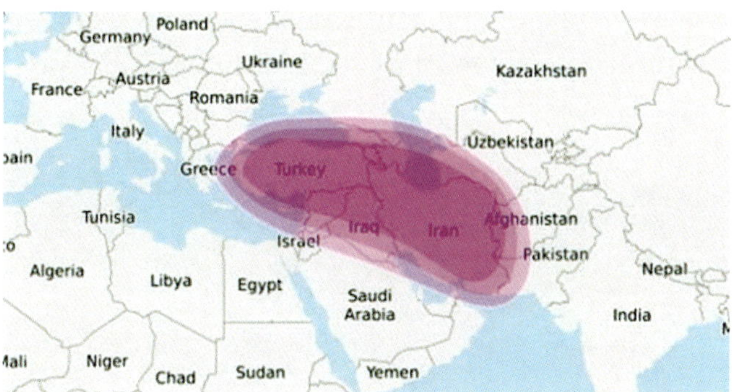

Figure 24. Common Ethnicities of Zaza Candidates – MyHeritageDNA

During the Old Sumerian period (circa 2900-2300 BC), Sumerian civilization was divided into several states, with one city as the centre of each state. It seems that the priesthood controlled much of what occurred in these states. The city states were often at war with each other over land areas. The early civilization of the Hittites (around 2500 BC) dominated most of ancient Eastern Anatolia. Clay tablets found in **Hattuşaş (Boğazkale)** are written in eight different languages, illustrating the prominent role that this city and civilization played in international travel during its reign.

DNA is the tree of the unwritten ethnic groups. Effective isolation has given the Zaza people their own DNA. It shows that exclusivity and endogamy in Zazadiyar has made the DNA of the Zazas a distinctive species. DNA, languages, migrations, religious environments, living habits, agricultures and cultural activities are all pieces of the puzzle that contribute to our knowledge of the ancient Zaza people. The results of the 3,500 Zaza candidates in the Haplogroup clearly prove that the ancestors of the Zaza people belong to the ancient skeletal finds in Kermanshah and Mahabad about 6,500 years ago. The reasoning around this offers us a creative way to research our ancient ancestry. Thus, with some caution, the ethnic origin of the Zaza people can be distributed as follows: **about 65% Elamite**, **about 20% Sumerian**, **about 10% Hittite** and **about 5% Caucasian**. It is important to note, however, that these estimates are based on probability calculations.

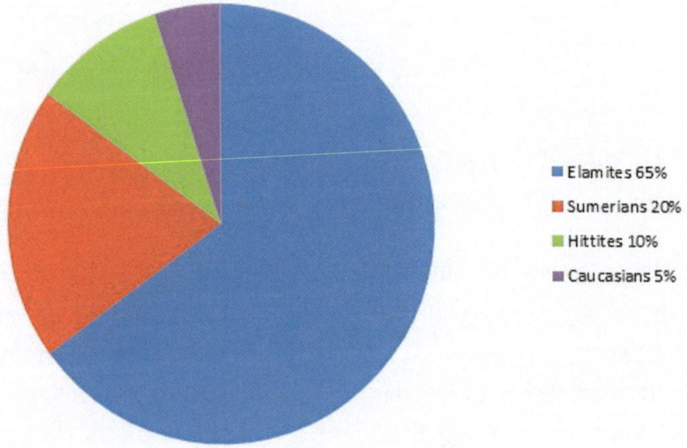

Figure 25. Ethnic Origin of 3,500 Zaza People

THE ALEVI RELIGION

Each Indigenous people have their own history of faith development. Parsist, Zoroastrian, Sumerian and ancient Persian mystical cults all left their mark on Alevism. Alevism originated in Anatolia among the Zaza people and is based on Zoroastrian, Sumerian and Shia Islamic foundations. The prehistoric heartland of the Zaza people in Eastern Anatolia includes the provinces of Tunceli, Bingöl, Muş, Kemah in Erzincan, the northern part of Elazığ and the southern part of Erzurum. According to Sumerian sources, in 3000 BC, the area was called **"Subartis"** and its inhabitants **"Suba"** in Sumerian. The provinces of the mountainous **"Subartis"** include Erzincan, Erzurum, Ağrı, Iğdır, Kars, Tunceli, Bingöl, Muş and Bitlis.

According to some sources, the word "Zaza" is mentioned in Sumerian cuneiform clay tablets (3100 BC). What this refers to may never be known to posterity. The Behistun Inscriptions of Iran, written between 515 BC and 486 BC in Old Persian, Elamite and Babylonian, by order of the Achaemenid Emperor Darius I, states that the Zaza people lived in a region northwest of Lake Van. According to several undocumented claims, the original home of the Zaza lies in an area near the sources of the Euphrates and Tigris rivers.

To a great extent, historical records such as the ***Bundahishn*** (circa 5th century BC), one of the sacred scriptures of Zoroastrian followers, show that the homeland of the Zazas can be located in the area around the sources of the Euphrates and Tigris, as it is today.

This indicates that the Zaza migrated to Khorasan in Northern Iran and not the other way around as previously thought.

The Sumerian word **"Za"** has to do with water, especially melt water. The word **"ab-za"** in Sumerian is an expression for world ocean. The snow supplies the Euphrates and Tigris with water. Bodies of water, spring floods and floods are associated with the Euphrates and Tigris rivers. The word Euphrates itself means "abundance". The river Euphrates is called in Old Persian **Ufratu**, in Elamite **Ú-ip-ra-tu-iš** and in Sumerian **Bu-ra-nu-na**. The Euphrates River has occasionally flooded Sumerian cities and roamed among its monuments throughout the ages. The Zaza people's land of origin "Zazadiyar" is at the headwaters of the Euphrates and Tigris.

From that aspect, if this conjecture is correct, it can therefore be taken as true that the word "zaza" refers to Indigenous people living at these source flows. Sumerian is very rich in doubling the noun. Examples of this are **du-du** = the words, **kur-kur** = the mountains, **šu-šu** = the hands, **küren-küren** = countries and so on. If this is the case, the word "za-za" can get its key as "The Flood People" at the land around the source of the Euphrates. Of course, this is a series of guesses and probability calculations, but can possibly help make these long stretches of time less desolate, more tangible.

Figure 26. The Emblem of Alevism

The Zaza people had the world's oldest monolithic religion, Zoroastrianism (1500 BC). The Zaza people converted between the years 669 and 744 to Islam and the conversion has taken place gradually. According to an unproven claim, the first province affected would have been Muş. Today, about 70% of the Zaza people profess the Alevism sect of Shia Islam and about 30% have converted to Sunni Islam. Dımılî-Zazas (South Zazas) were deported in the 1510s from the province of Tunceli and converted to Sunni Islam. Within Alevism in the Middle East there are three branches of faith Caferî, İsmailî and Fatimî. The Alevis in Anatolia belong to the Caferî communion. The word Alevism derives from the first Imam and the fourth Caliph Ali who was the son-in-law and cousin of the Prophet Muhammad. Followers of Alevism in Turkey are called **"Alevi"**, the name represents a Turkish form of the word Imam Ali, and is an elliptical compound word of **"Ali Evi"** (House of Ali), thus the term Alevi appeared only in the middle of the 17th century. In Turkish, double vowels and double consonants are avoided in a compound word, that is, the **"i"** falls away, which means that **"Ali Evi"** becomes **Alevi**. For example, elliptical words in Turkish are Kahve Altı = Kahvaltı, Teli Oğlu = Teloğlu, Maşa Allah = Maşallah, İlla Allah = İllallah, Dıl Lav = Dılav and so on.

Alevism has a heterodox structure and has a liberal view of cult development, unlike the orthodox view. Alevism does not strictly follow formal rules. In other words, it has a contemporary and humanistic feature that constantly renews itself through the filter of the mind and logic and keeps up with the development of technology and science. The bloody events of oppression from the past have consolidated their development until our time, but have changed and transformed from old to new, from negative to positive. During this process, the Alevis continued their path by never allowing dogma or superstition, and sometimes by breaking the insurmountable taboos that had been imposed.

The most important of the special religious characteristics of Alevism are that Alevis do not pray five times a day, do not fast during Ramadan, do not have to give alms or make pilgrimages to Mecca. **Hıdırellez** is the Alevi fasting period of 3 days in February and 12 days in the month of **"Muharrem"**, the month of January in the **"Hijri"** almanac. **Cemevi** is their house of prayer and meeting place. Women in the Alevi society have a relatively high status. Women are equal, there is no requirement to wear a veil, men and women participate together in cult ceremonies, and alcohol is consumed in connection with initiation rites, and finally, Sharia is not strictly enforced.

Throughout history, the fate of the original inhabitants of Eastern Anatolia has always been overshadowed by constant forced relocations ordered by the ruling élite. Migrations are thus a phenomenon among the Zaza people from the 13th century until the present day. From the heartland, Zazas emigrated and were deported to different parts of Turkey. It had the consequence of loss of the mother tongue and conversion to Sunni Islam. Upon emigration to Southeastern Anatolia, Diyarbakır, Mardin and Urfa, these groups were converted to Sunni Islam, but retained the native Zazaic language – they became ***Sunni Zazas***. The Alevis who were deported and emigrated to Sivas, Kayseri, Adana, the slopes of the Taurus Mountains (Tahtacılar), Tokat, Amasya and Çorum lost their mother tongue – they became ***Alevi Turks***. The Alevi group that emigrated to Maraş, Malatya and Adıyaman got their mother tongue, Zaza mixed with the Kurdish language – they became ***Alevi Kurds***. The Ömeran Zaza tribe was deported from Tunceli and Bingöl provinces to Diyarbakır and Mardin provinces and migrated after 1850 to Konya district. They converted to Sunni Islam, but have retained features of the Zaza dialect – however, they identify themselves as ***Kurds***. Small groups of Zaza tribes that have also spread in patches have converted to Christianity – they exist as ***Christian Zazas***. In this manner, several groups can coexist in a Zaza tribe – ***Alevi Zazas***, ***Alevi Kurds***, ***Alevi Turks***, ***Sunni Zazas*** and ***Sunni Kurds***.

Alevi Zazas generally live in the provinces of Tunceli, Erzincan, Bingöl, Sivas, Gümüşhane, Muş, Erzurum, Maraş, Bayburt, Malatya, Kayseri, Ardahan, Kars, Tokat, Yozgat, Çorum, Adıyaman, Elazığ and Amasya. Sunni Zazas, on the other hand, mostly live in the provinces of Bingöl, Diyabakır, Mardin, Bitlis, Adıyaman, Urfa, Erzurum, Elazığ, Siirt, Van, Muş, Konya and Tunceli. Somehow, an Alevi in Erzincan and an Alevi in Balıkesir do not live in the same conditions, and the perception, content and function of Alevism vary from person to person.

Since the Kırmanç language has not been written down due to oppression and prohibitions, especially the younger generations have been rendered unable to speak and use their mother tongue. For these reasons, some of the people of Southern Dersim (Pertek, Çemişgezek, Mazgirt) have become Kurdish or Turkified. A significant portion of them also preserve their political identity.

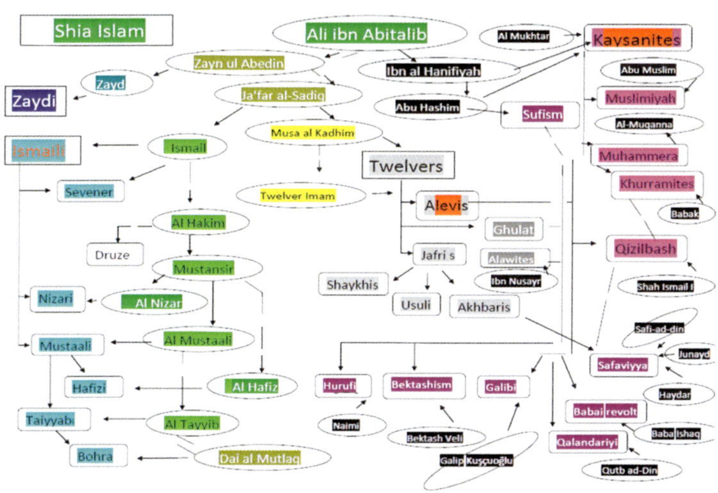

Figure 27. The Historical Emergence of the Alevi Tarikat

Estimated sect breakdowns of Alevi/Sunni within the three Zaza groups are estimated as follows: Kırmanç Zazas (North Zazas) are about 95% Alevi and 5% Sunni, Kırdkî Zazas (Central Zazas) are about 75% Alevi and 25% Sunni and Dımılî Zazas (South Zazas) are about 80% Sunni and 20% Alevi. Kırdkî Zazas in Bingöl, Elazığ and Sivas identify themselves as Kırdaşkî Kurds.

The Alevis in Turkey should not be confused with the Arabic-speaking **Alevites** (Alawites) "Nusayris" in Syria. Both Alevis and Alevites have in common their worship of the first Imam Ali. The common point for both groups is their love of **Ehlî-Beyt** (Ahl al-Bayt). Ehlî-Beyt means "House of the Prophet", which refers to five family members of the Prophet Muhammad, including himself (died 632), Imam Ali (died 661), Hazrat Fatima (died 632), Imam Hasan (died 670) and Imam Hüseyin (died 681).

Alevites "Nusayris" in Syria are characterised by a variety of habits and customs that are contrary to Orthodox Islam, such as wine drinking, use of incense, and their doctrine of reincarnation and belief in the Trinity of **Muhammad**, **Ali** and **Salman**. Alevites "Nusayris" have mosques and pray five times a day and participate in the fasting month of Ramadan. They neither have *Mürşit*, *Pir*, nor *Rayber*. There was never the practice of polygamous marriage in the history of the Alevi Zazas. Some Kurdish Alevis in Turkey celebrate the Fire Festival (Nowruz or New Year). The Anatolian Alevis do not eat rabbit, crustaceans or birds of prey, which are ***haram*** (forbidden). Neither men nor women typically shave their pubic hair, but men in the military must do so.

The Anatolian Alevis are Zazas, Turks and Kurds. Most Alevis live in Turkey, where they are a minority, while Sunni Muslims are the majority. Alevi communities are concentrated in Central Anatolia, in a belt from Çorum in the west to Muş in the east. Due to persistent harassment, there are many Alevis who live as crypto-Sunni Muslims outside the core region. The size of the Alevi population is similarly questioned, but most calculations estimate it at somewhere between 25 and 30 million people or about 30% of the total population of the country. Estimates of the proportion of Turkey's population that are Alevis typically range from 20-30% to as much as 40%. Scattered minorities live in Bulgaria, Albania (Bektaşî), Macedonia, Kosovo, Iran, Iraq, Lebanon, Cyprus, the Caucasus, Greece and the Turkish diaspora (mostly Western Europe and the United States).

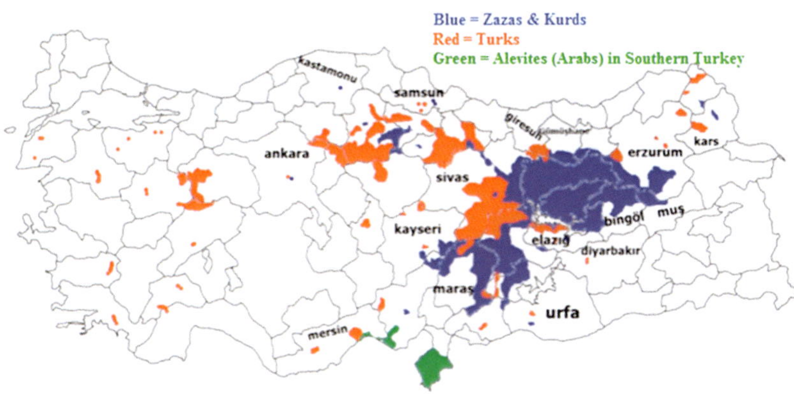

Figure 28. Distribution of Alevi Population in Turkey

In the past, Alevis were grouped as **Kızılbaş** (Redheads), a generic term used by Sunni Muslims in the Ottoman Empire for the various Shia sects of the 1510s during Yavuz Sultan Selim's reign. The name Kızılbaş first appeared in the 15th century. Many other names exist, often for subgroups, including Tahtacı, Çepni, Abdal and Amucalı. The Kızılbaş and the Bektaşi Order shared common religious beliefs and practices and became intermingled as Alevis despite many local variations. Isolated from both the Sunni Ottomans and the Twelver Imams Shia Safavids in Iran, the Alevis developed traditions, practices and doctrines by the early 17th century that marked them as a closed autonomous religious community. However, there are Alevi philosophies, customs and rituals that are noticeably different from those of the Twelver Imams Shias in Iran, Iraq and Lebanon. As a result of immense pressure to conform to Sunni Islam, the Alevis developed a tradition of opposition to all forms of external religion.

During the centuries of the Ottoman Empire, Alevis were prohibited from proselytising and Alevism regenerated itself internally by paternal ancestry. To prevent the invasion of hostile outsiders, the Alevis insisted on strict endogamy, which eventually made them a quasi-ethnic group. Some of the differences that mark Alevis from Sunnis are that they regard themselves as true Muslim followers of Hacı Bektaş Veli who emphasise the role of the Imam Ali in addition to the oneness of God and the prophecy of Prophet Muhammad. Alevis accept the Imam Ali as the only legitimate successor to Prophet Muhammad and add to the witness formula **"Kelime-î şehadet"** the words "and the Imam Ali is the friend (or vicegerent) of God". For Alevis, the Prophet Muhammad and the Imam Ali are both emanations of the *Divine Light*, Prophet Muhammad is the announcer and the Imam Ali is the preserver.

Alevis and Islam

The Islam of the Anatolian periphery encountered in the Kızılbaş-Alevi tradition comprises a unique set of religious practices and ideas. What marks Alevism as Islamic is the Islamicate context of its historical formation and its terminology, which is markedly Sufi and Twelver Shiite. Alevis, as well as Bektaşis, venerate the first Shiite Imam Ali and his descendants. Like Twelver Shiites, Alevis remember their own history as a history of suffering, for which the tragedy of Karbala has remained the symbolic and emotional anchor point until today. According to the Shiite tradition, the third Imam Hüseyin, the sole legitimate leader of the Muslim community, and most of his closest relatives were massacred by Umayyad troops in the desert near the Iraqi settlement of Karbala in 680. Despite the importance of remembering the tragedy of Karbala, the Alevis have, however, little in common with the mainstream of Twelver Shiism.

Traditional Orientalist scholarship has argued that their creed has been affected by certain elements of ***Ghuluw*** ("exaggeration" in the veneration of the Imams) – a term used in early Shiite apologetics and heresiography to condemn individuals and groups as heretics. These apologetics contributed to the formation of Shiite orthodoxy. The veneration of Imam Ali as divine, the idea of the incarnation of the divinity in human form **(Hulul)** and the idea of the transmigration of souls **(Tanasukh)** were branded as **Ghuluw**. For Alevis (as well as Bektaşis), God is not only transcendent, but also manifest in the immanent world, especially in humans. The Sharia plays at best a minor role in this system. It is interpreted – in particular by Alevis – in an allegoric sense, as an initial step on the mystical path that is of only limited importance for the initiated. Mainstream Islamic authorities regularly condemned such interpretations and the resulting negligence of Sharia prescriptions in practice as unacceptable "libertinism". For both Bektaşis and Alevis, Sharia has a different meaning. It is integrated into their interpretation of the ***Tasawwuf*** teaching of the "four gates and forty stations" **(dört kapı, kırk makam)** that functions as a metaphor for and normative orientation on the mystical path. The four gates and the forty stations (ten per gate) associated with them consists of ethical and spiritual rules that structure that path. They advance from more practical and social responsibilities (the first gates of ***Sharia*** and ***Tarikat***) to the spiritual and mystical stages of the third and fourth gates (***Marifet*** and ***Hakikat***). The goal of the mystical path is to advance to the last gate and turn into a "perfect human being" (insan-î kâmil).

The central ritual of the Alevis, as well as the Bektaşis, is the ***Ayin-î Cem*** (ceremony of unity). While the remembrance of Shiite mythological events has a prominent place in the ritual, another sequence of the Ayin-î Cem celebration re-enacts the mythical "Ayin-î Cem of the Forty" **(Kırklar Ayin-î Cemi)**, in which forty saints participated. It took place in mythical time, during Prophet Muhammad's nocturnal ascension to the heavens, and is interpreted by the Alevis as allegoric proof of Imam Ali's spiritual superiority over Prophet Muhammad. The Ayin-î Cem ritual is structured by lamenting hymns that are accompanied by the weeping sound of the ***Saz*** (long-necked lute), the traditional musical instrument of the Alevis. As in the mythical "Ayin-î Cem of the Forty", men and women celebrate the ***Ayin-î Cem*** together and join the ***Semah*** ritual dance. This has led outsiders, for whom participation in the ritual used to be strictly forbidden, to the false notion that Alevis conducted wild orgies.

Figure 29. Ayin-î Cem Gathering – Eastern Anatolia

The Background of Alevism

Historians of religion often describe Alevism as syncretism. Syncretism refers to processes by which practices and ideas formerly independent of each other are drawn together in a new form. Thus, the concept focuses not on the continuity of religions, but on their dynamic nature, in response to changing environments. It carries, however, the risk of an implicit hierarchization of different religions according to the extent of their mixing and the respective degree of "purity". Engrained in the concept is an implicit conceptualization of religions as being distinct from each other and their respective "non-religious" environments, as being differentiated and bounded entities with clearly traceable origins. Such normative assumptions, combined with the insight that every historical religion as such is necessarily a product of mingling and differentiation processes of various earlier cultural elements, put the heuristic usefulness of the concept into doubt.

The socio-religious traditions known since the early twentieth century under the label Alevism share practices and beliefs that bear points of contact with charismatic Sufism, a Shiite mythological orientation and Turkish, Kurdish-Iranian, as well as Anatolian non-Islamic cultural and religious traditions. Over the course of the twentieth century, these traditions were reinterpreted in the light of modern ideals and values. The Alevism that emerges as a product of such processes of reinterpretation may look very different from imaginations of "traditional Alevism" but is certainly no less real and authentic. The notion of a "traditional Alevism" as the core tradition of Alevism is itself a product of modern longings for authenticity and clearly definable origins. Like other living religious traditions, Alevism should not be understood as a fixed entity with clearly definable origins and boundaries, but as a set of historical traditions, constantly adjusting and rejuvenating themselves, thereby drawing on a multitude of sources and inspirations and in dynamic contact with their various environments.

Pragmatically, one may regard the Alevi tradition as part of the Islamic discourse, to the extent that the carriers of this tradition relate themselves to Islam and are regarded by Muslims as part of the Islamic fold. Even the accusation of heresy marks the Alevis as part of the Islamic tradition, since someone accused of transgressing a religion's limits is a priori perceived as being part of it. The way in which the discussion about Alevism in relation to Islam is led is in itself very modern. It is concerned with origins along imaginations of national, ethnic and religious purity, postulates singularity and historical continuity, and emphasizes clear demarcations between different religious formations.

Processes of objectifying religion have long been present in the Ottoman Empire, intensifying since the sixteenth century when the Ottomans began to more explicitly turn to Sunni Islam as the basis of their religious authority. It is in the sixteenth century that we find in Ottoman administrative and court records many examples of the accusation of heresy directed against the Kızılbaş and other non-Sunni groups. Such accusations contributed to the historical constitution of a Sunni-Ottoman identity. In the Turkish Republic, too, the Turkish national subject has often been marked as Sunni and Turkish, thus positioned against Alevis and Kurds. Hence, in the state-centric, nationalist discourse of Turkey, similar to the way in which the demands of the Kurds for ethnic recognition and autonomy can be interpreted as ethnic separatism, the call of the Alevis for the recognition of their difference from Sunni Islam is perceived as dividing the nation by some of those participating in the nationalist discourse. Both Kurdish and Alevi identities challenge homogenizing ethnic and religious interpretations of Turkish nationhood in the Kemalist tradition of "Unity and Togetherness" **(Birlik ve Beraberlik)**. This hegemonic religious discourse of Turkey presents the Alevis with jour main alternatives to strengthen their position.

Alevis today face the immense task of maintaining a pluralist tradition while adapting to a political context that strives for religious homogeneity through assimilation/return of the Alevis into the Sunni mainstream. They often find themselves in very ambivalent situations, for example, when they, for the sake of preserving their tradition, embark on Dede/Seyit training programmes, even though the homogenization that this entails harbours the danger of further undermining traditional Alevi plurality. Reservations among Alevis also exist with regard to the efforts by Alevi umbrella organizations toward a certain standardization of Alevi practices. While Alevis are unanimous in their principal demand for acknowledgement of their difference from Sunni Islam, the political, religious and cultural differences among the various Alevi currents make it difficult for them to unite. Therefore, the plurality within the Alevi tradition is likely to endure, even if the grounds for inner-Alevi pluralism have changed considerably. Today, dividing lines are based less on lineage and traditional affiliations than on political and cultural preferences, economic opportunities, as well as practical considerations.

Spiritual Leaders and the Worship

Intellectual Alevis and spiritual leaders in different communities strive to preserve and define the Alevi identity, tradition and group history. They have their own religious ceremonies known as **"Ayin-î Cem"**, led by spiritual leaders who are called **"Dede/Seyit"**, who inherit their profession. Dede or Seyit is another collective way for ***Mürşit***, ***Pir***, or ***Rayber***. The word "Dede" is actually an exonym and is the Turkish name. At meetings, religious poetry, **"Deyiş"** is read, Turkish/Zaza Alevi songs are sung in a rhythmic manner, and women and men dance ritual **"Semah"** together. Alevis have an extensive and widely read religious literature, composed mainly of spiritual songs, Deyiş poems and epic verse. Semah is performed by men and women together, to the accompaniment of **"Saz"**. The dancers, for example, symbolise the revolution of the planets around the sun by men and women turning in circles, and by standing themselves and uniting with God. It is no coincidence that the cosmic dance, with choreography based on the rotation of planets, stars, star clusters, and galaxies around their own shoulders and around larger systems, is also "Semah", which means heaven. However, there are many regional variants of Semah.

Figure 30. The Semah Ritual – Western Anatolia

The central Alevi communal worship service is called a **"Ayin-î Cem"** (***Ayin-î Cem*** or ***Cem***, meaning amass, gathering, congregation or assembly meeting), which is performed in special buildings called **"Cemevi"** (House of Gathering). Alevis believe that Ayin-î Cem has its roots in the original worship and teaching meeting of 40 spiritual individuals, **"Kırklar Meclisi"** (Gathering of the Forty)**,** led by the Imam Ali. It takes place in a Cemevi. The supposed prototype of the ceremony is the Prophet Muhammad's nocturnal ascent into heaven, where he beheld a gathering of forty saints **(Kırklar Meclisi)**, and the divine reality made manifest in their leader, the Imam Ali. During the "Ayin-î Cem Ceremony", the troubadour **"aşık/zakir"** plays the "saz" while singing spiritual songs, some of which are centuries old and well-known among Alevis. Each song, called a **"Nefes",** has a spiritual meaning and aims to teach participants important lessons. For example, one such song states: "Learn from your mistakes and be knowledgeable. Don't look for faults in others. Look at 73 different people in the same way. God loves and created them all, so don't say anything against them". The love of the creator for the created and vice versa is symbolised in the Ayin-î Cem ceremony by the use of fruit juice or sorbet, which represents the intoxication of the lover in the beloved. At the closing of the Ayin-î Cem ceremony, the person leading the ceremony engages the participants in a discussion, which is called a **"Sohbet"**.

Musahiplik

"Musahiplik" (being best man) is a covenant relationship between two men of the same age, preferably along with their wives. The word **"Musahip"** is "best man" in English. In a ceremony in the presence of a Dede, the partners make a lifelong commitment to care for the spiritual, emotional and physical needs of each other and their children. The ties between couples who have made this commitment are at least as strong as those between blood relatives, so much so that "Musahiplik" is often called a spiritual brotherhood **(manevî kardeşlik)**. The children of covenanted couples are not permitted marry each other as if they shared a blood relationship.

The Tahtacı Alevis identify "Musahiplik" with the first gate **(şeriat)**, since they regard it as a precondition for the second **(tarikat)**. Those who attain to the third gate (**marifet**, "gnosis") must have been in a Musahiplik relationship for at least 12 years. Entry into the third gate dissolves the Musahiplik relationship (which otherwise persists until death) in a ceremony called "Öz Verme Ayini" (the ceremony of giving up the self). The value corresponding to the second gate (and necessary to enter the third) is "aşinalık" ("intimacy", perhaps with God). Its counterpart for the third gate is called "peşinelik", for the fourth port (**hakikat**, Ultimate Truth), "cıngıldaşlık" or "cengildeşlik".

Kirvelik

"Kirvelik" (sponsorship) is a form of ritual co-parenthood practised in Eastern Anatolia and established through the Islamic ritual of circumcision. **"Kirve"** is the sponsor of the boy by circumcision. It has no formal base in religious law. However, it is known from pre-Islamic poetry that the custom of circumcision existed in early Arabia. With respect to religion, both members of various Alevi-Shia groups and Sunni Muslims practise "kirvelik". Among the Alevis (Zazaic, Turkish and Kurdish), this practice is especially appreciated. Circumcision ceremonies are celebrated with pomp and parade. Entertainment is organised, people dance the chain dance "halay", they shoot with weapons, they compete in javelin throwing on horses "cirit" and so on. The boy is always dressed in a white military uniform or prince costume, a "baby-general". According to the Alevi tradition, the Prophet Muhammad placed great importance on this ritual relationship when his grandsons Hasan and Hüseyin (the sons of Imam Ali) were circumcised. Therefore, the Alevis believe that the blood of the Imams (Hasan and Hüseyin) passes between the sponsored and the sponsor.

The "kirvelik" involves three central elements: 1) the rite of passage associated with the circumcision of a male child (who is usually between 7 and 12 years old): 2) the socio-spiritual ties established between the participants: and 3) sponsorship. The "kirvelik" relationship has several basic characteristics. It is very much like kinship, but it is formed voluntarily. The relationship between the two families is one of trust, mutual assistance, close friendship and respect. The rights, duties and responsibilities between the sponsor and the child parallel those between father and son. The sponsor shares responsibility for the boy's circumcision expenses, training, education, well-being and marriage. In return, the boy reciprocates with the loyalty, obedience, respect and affection he gives his own father. People claim that it is more shameful to disobey or disrespect one's "kirve" than one's father. The incest taboo is extended to both families.

A further example of reciprocity involving "kirve" from the Zaza society is connected with business activities and demonstrates the deep feelings of fellowship and solidarity the partners have towards each other. This is an ancient tradition but is still alive particularly in the villages. For example, when a person starts a business he will borrow interest-free money from the "kirve", a friend, a relative or a fellow villager. The donors feel obliged to contribute, especially in starting up a business. The obligation is as if one is helping his own brother. That the money is given free of interest is due to the status and esteem this provides for the donor. There is also an implicit expectation that the donor might end up in a similar situation and need help from the receiver. This transaction of money may seem to be purely economic but it also involves other humanitarian considerations such as compassion, conscience, sense of duty, avoiding a loss of honour and so on.

However not everyone is willing to reciprocate. The poor or economically weak cannot always reciprocate and as a result, they must remain indebted to each other, and naturally not everyone wants to be in this situation. Poor people cannot afford to arrange large feasts and will have a limited ritual. While the sponsored boy can marry his cousin, he must neither marry nor have sexual relations with the offspring of the sponsor.

Kirvelik and Musahiplik – Alevis and Sunnis

Within the borders of Tunceli, it estimates that Alevis constitute more than 90 per cent of the population. In other words, the Alevis are "in power". Nonetheless, this power emanates from being the majority in demographic terms, not in administrative terms. The unequal demographic distribution of Alevis and Sunnis has led to relations of power that have, throughout history, varied in form and content. In its view, both Alevis and Sunnis have taken advantage of some religious institutions or authorities for their economic, political and safety-related benefits. As a matter of course, Sunnis living among the Alevi majority sought the advantages of possessing strong patronage relationships with Alevi families and the help of Alevi religious authorities, and much more. In this regard, when looking at the nature of such power relations, it can speak of two different periods – before and acer the 1970s, when Alevis and Sunnis became politically polarised. Nevertheless, some Alevi institutions seemed to be presented as conflict-solving mechanisms, until the politicisation of religious identities.

In this sense, the "kirvelik" institution gains noticeable importance. For example, by forming "kirvelik" bonds with an Alevi family of high economic or social status, Sunnis could gain a strong foothold in the community. Due to the strong and diverse social ties existing among Alevi communities, the scope of these "kirvelik" ties would eventually expand. All Alevi families are connected through patrilineal bonds to larger families in terms of the ***Aşiret*** (tribe). These extended families are in contact with a large number of other families through the ties of **Kirvelik** or **Musahiplik**. In addition to such attachments, the multi-dimensional and complex tribal and **Talip-Pir** relations of Alevis were also factors determining the range of their social relationships. When we consider that all these relations existed in the close vicinity of places inhabited by Alevis, we can see that Sunnis' "kirvelik" ties greatly enriched their web of social relationships with the majority. Religious faith is both the source of the ***Kirvelik*** and ***Musahiplik*** relations and the basis of their enforcement. Therefore, the social processes emanating from these relations bore a certain consistency, founded on religion. Because of this, the influence that the Sunnis gained through the "kirvelik" institution was significant for them.

The Sunnis living in the villages and hamlets of Pertek have very much adopted the concepts dominant in the social and religious lives of the Alevi majority. The reason for this is that they have shared this land, where production is based on agriculture and non-nomadic farming, and all its natural resources (such as forests, rivers, pastures and fields) for centuries. Mutual problems independent of human activity (such as natural disasters) affecting these sources, as well as shared difficulties experienced in the markets where the product gains its exchange value, have brought the two communities together on the mutual basis of "being members of the peasantry". This has resulted in the emergence of a common attitude and behaviour patterns regarding various problems. Perhaps the unique examples of these relations are the shared beliefs and rituals related to the existence of various practices of ***Ziyaret***, sacred to both Alevis and Sunnis. Furthermore, narratives of common religious-social activities sometimes performed under the guidance of ***Pirs***, and sometimes ***Raybers***, exist among Sunnis.

Until the 1970s, mutual contact between the communities was facilitated through the ***Ayin-î Cem***, the basis of Alevi religious structure, with the **Talip-Pir** institution, on one hand, and the Sunni community leaders, known as "prominent families", on the other. Of course, other religious figures were just as important as the **Raybers**, who were prominent families in Sunni communities. The ***Pirs*** and the

Raybers preserved the legitimacy of inter-community competition and communication for centuries. The defects in the Turkish Republic's economic and social structure also contributed to the continuation of such feudal actors and institutions.

Additionally, Sunnis did not opt to take advantage of the state institutions. This can also be understood as an indicator of their lack of trust towards the Turkish Republic. Considering the state's view of Turkishness and Sunnism as its founding elements, such an approach would have been wise for the Sunnis. However, until the confrontations organized by the state, the Sunnis displayed no such tendency. The reason may have been that Dersim constitutes a region that, despite the constant reform policies put into practice both by the Ottoman Empire and the Republic of Turkey, has maintained its Alevi majority and identity. Thus, one may claim that the general interactions of the Sunni community have been mainly with the Alevi majority. Sharing a common social, economic and religious space with the Alevis has led the Sunnis to accept the majority's authority and institutions.

Even though traditional relations have lost their relevance, due to the dissolving consequences of modernization and politicization influencing socioreligious relations, "kirvelik" has maintained its importance. Alevi customary law still organises and sustains local daily like. Evidence shows that Alevi law has gained legitimacy in areas populated by Sunni majorities, such as the district centre of Pertek. Even today, for many Sunni families "kirvelik" constitutes the basis of their inhabitation in rural settlements in Pertek and the surrounding areas.

Apart from "kirvelik" relations, Sunni communities have developed another strategy around the socio-religious institutions (Ayin-î Cem rituals) of the Alevi majority, which dominated daily life and were strongly re-consolidated in the Republican era (after 1938). In this way, they could pressure the problematic individuals or groups through social and religious institutions whose effectiveness far surpassed that of the state. Until this period, Alevi communities' internal issues, especially those in rural areas, were never resolved through state institutions. According to an elderly Alevi interviewee, during the Ottoman period, applying to a ***Şeriat*** (Sharia) court for the resolution of an internal issue was seen as such a heinous crime by Alevis that it was punishable by ***Aforoz*** (excommunication). Therefore, in an environment where most of the population refused to acknowledge the state as a tool of reference for resolving intra-community matters, the Sunni minority applied to those institutions that held power over the members of the majority.

For the Sunnis of Pertek, the Alevi majority's dominance has been more limited when compared to the countryside, where the Sunni communities rely on the Alevis' social structure. The relationship between Alevis and Sunnis in the countryside reflects the social institutions of the Alevi majority. In contrast, the Sunni-dominated official institutions and the dominant traditional structure determine commercial and social life in the district centres. This spatial difference came to the fore in various forms in the 1970s.

The Tombstone Visit

In the event of a death, relatives and acquaintances gather three days after the funeral to commemorate and honour the memory of the deceased. Close family members carry grief for 40 days as a sign of respect. Male family members do not shave for 40 days. Nowadays, participants (in optional clothing) are offered food (meat dishes) on the day of the funeral and 40 days after the funeral. Before the grave of the deceased is closed, some money is put on the funeral. These coins are divided among the hodja and the gravediggers. In the Alevi Zazas there is a tradition of visiting the cemetery and handing out food in the spring, which is called ***Mezar Kaldırma*** "Erection of Tombstone". This is done with the belief that it will be for the benefit of the deceased. They erect the tombstones of their relatives who died in winter and autumn at the beginning of summer. On a day that is usually decided in the first week of June, men, women, children and all the people of the village go to the cemetery, wearing their most beautiful clothes. They take meat from the animals they sacrifice, dishes, donuts, cookies, tidbits and halvas with them. Everyone collapses on the graves of their relatives leaves flowers, kisses the tombstones and pray in tears. At the end of the ceremony, the villagers distribute the food, cookies and tidbits they bring with them to the people and children and make their relatives read their past. When the food has been eaten together, everyone says goodbye by kissing their relatives' tombstones and returns to the village in the same way. After the ceremony of setting up the tombstone, the deceased's wife or husband could get married.

The Tribute and the Talip

Within Alevism, there is a three-man hierarchy, with Mürşit being the high priest, Pir receiving an order from Mürşit and Rayber receiving an order from Pir. Once a year, the spiritual leaders turn up a Mürşit, a Pir or a Rayber (not simultaneously) to collect tributes (bounties) in Zazaic ***Çıralık*** and in Turkish ***Hakkullah***. Then they wash his feet (without soap) and drink this holy water to stay healthy in the future. The tribute (Çıralık) can be in the form of money, house pets, or goods. The goods cover everything from cigarette lighters to vehicles. Their area of activity for exploitation stretches over the entire Alevi region in Eastern Anatolia and the Alevi ghetto **(gecekondu)** concentrations in the large Turkish cities. Some of them also travel among the Alevi milieus in Western Europe. From the Alevis in Western Turkey and Europe, they collect only money, goods and clothes. The Pirs/-Raybers divide the territories among themselves. Consequently, each Pir/Rayber has his own territory where he is engaged in his activities.

Spiritual leadership is inherited, passed from father to son. The rhetoric of the Alevis about gender equality is generally that only men are allowed to fill such leadership roles. The tribute payers go by the name of "Talip/Mürit" and the spiritual tribute recipients "Dede/-Seyit". Each Dede has a number of Talip families. Ordinary Alevis (the Talip people) would owe a certain Dede lineage (but not others) on the basis of existing family relationships. At the ceremony, receiving tribute, the Talip family stands in **"Peymançe"** position to leave "Çıralık" to Dede, in connection with this everyone should kiss his feet and hand. A Dede can, of his own accord, request what form of Çıralık (money, goods, or house pets) should be and in such circumstances, the Talip individual should not protest. In all forms, a Dede/Seyit is supernatural and regardless of age, it is concluded that all Talip/Mürit individuals (human figure) should show great respect by kissing his hand. In principle, Dede families (Mürşit, Pir, or Rayber) do not own any estates, but they can have livestock. Their only source of income is tributes (Çıralık) from their Talip families. Marriage between Dede/Seyit and Talip/Mürit is strictly forbidden.

Figure 31. Eastern Anatolia 1200 BC

In the wake of urbanisation from the 1980s onwards and socialist influence, which looked at the Dede people with suspicion, the old hierarchy has largely been broken down. The practice of tributes (bounties) has largely been abandoned. Many Dedes now receive salaries from Alevi cultural centres, which undoubtedly subordinate them. Such centres no longer have social enterprises or consultations, such as the old reconciliation ritual, but they emphasise the performance elements of the music (Saz) and dance (Semah). The Dedes are now, contacted on a voluntary basis and their role has become more circumscribed, that is, limited to religious rituals, research and counselling.

Offspring of the Prophet

Of the approximately 220 Zaza tribes, 12 are tribes of the **"Evladî-Resul"** (Offspring of the Prophet). According to the legends, Evladî-Resul tribes are descendants of the Prophet Muhammad. However, it worth noting that there is no historically reliable evidence for these legends. These 12 Dede tribes ***Ocak*** in Turkish are Kureyş, Baba Mansur, Derviş Cemal, Ağu İçen, Sarı Saltık, Dede Karkın, Hıdır Abdal, Şeyh Ahmed, Üryan Hızır, Sinemilliler, Pir Sultan and Seyit Sabun. All these 12 tribes are represented in DNA tests in the summary **"Zaza People's Origin Based on DNA"** (2021). No differences have been found in ethnicity estimates between the Zazas and the tribes of Evladî-Resul. Traces of being descendants of the Prophet Muhammad have not been discovered with DNA testing. Admittedly, this is somewhat disappointing as I had hoped for a clear result.

The **"Quraysh"** (Kureyş) is a Bedouin group from Northern Arabia in the 400s. The Prophet Muhammad belonged to one of the lesser clans of Quraysh, had no powerful lineage and no wealth of his own. If there is a connection between the Prophet Muhammad's clan Quraysh (Kureyş) and the Zaza tribe Kureyş, posterity will never know as there is no solid evidence for the claim. The old beliefs have been absorbed into history with mythical facts. The descent of the Dede tribes is, on the whole, a questionable assumption. Thus, it is time to draw conclusions and switch track, as the historical scenarios provide no clear answers and provenance from the Imam Ali is questioned. For almost 1,400 years, the Alevis (Talip people) have paid large sums, drank the feet-washing water, kissed their feet and for good reason have been made to believe that they were descendants of the Prophet Muhammad. Everyone in the Middle East is a descendant of the Prophet Muhammad, but no one carries his genome.

Inherited Ritual Acts and the Jâr

Many Sumerian ritual acts have deepened in the Zaza people. Examples include the Semah Ritual, the Twelver Employees, Amulets against Evil and so on. The common man had to keep to his private divinity and, through prayers and amulets, try to protect himself against accidents and evil as best he could. The Twelver Employees are as follows:

Figure 32. The Twelver Employees 2000 BC – Anatolia

Alevi (in a Cemevi)	**Sumerian** (in a Temple)
1. Pir	1. Driver
2. Rayber	2. Sheriff
3. Observer	3. Usher
4. Lighter	4. Warrior
5. Zakir (Aşık)	5. Musician
6. Cleaner	6. Bird Keeper
7. Cupbearer	7. Goat Herd
8. Waiter	8. Weir Inspector
9. Messenger	9. Messenger
10. Inviter	10. Grain Controller
11. Footman	11. Courtier
12. Porter	12. Porter

The above-listed **Zakir** – derived from the Arabic word **Zikr**, meaning "to remember, to hold in mind" – is the person who plays and sings the music at Alevi **Ayin-î Cems**. In some regions, he or she is known by other names as well, such as ***Aşık***, ***Sazcı***, ***Sazdar***, ***Sazandar***, ***Sazende***, ***Guyende*** and ***Kamber***. Some of the **Zakirs** (performers) are also local Aşık (poets) in their region. In some ***Ocaks*** (genealogical lines) and regions, **Dedes/Seyits** perform the music as well, with no separation from the Zakir. Some dries are also Aşıks. Therefore, Zakir defines the performer in **Ayin-î Cem** rituals, but Aşık defines those who write poetry and play and sing under their pseudonyms.

Whether performed by a recognised Aşık or only a Dede, Alevis consider the service of the Zakir to be a sacred duty and see Zakirs as "sacred Aşıks". They believe that Aşıks reflect all the aspects of life, including the religious aspect, in their poetry and constitute the most important transmitters of oral tradition. For Alevi-Bektasşis, Aşıks who play the ***Saz*** and write lyrics are considered to be the "wise men" and the "true dervishes" who leave a mark on their times.

The Zakirs' duty has been identified with different holy figures in different periods. The most critical Alevi written sources – the ***Buyruk*** writings – contain sections on the functioning of the Ayin-î Cem and the services, including that of the Zakir, and frequently emphasise that this duty is the legacy of the archangel Cebrail (Gabriel). Several other sources list the names of Imam Ali's sons, Abdu's-Samed and Hasan, as well as Bilal Habeshi and Imam Cafer, as the holders of the services in the "Ayin-î Cem of the Forty".

The sacred character of the Zakir position derives from its status as a carrier and transmitter of the holy words that symbolise the power of God. Because the Zakir performs this duty with his ***Saz***, this instrument is considered equally sacred.

The Zaza people have taken over many elements from Zoroastrianism. There is reason to believe that Zoroastrianism lives on as a kind of residue religion in Alevi Zazas. For example, the sunrise (the light) is considered holy, divine and inviolable. Eastern Anatolian Alevis believe that it is necessary to wake up before sunrise and that praying towards the sun when it hits the earth brings good luck. Alevi Zazas believe that the sun's rays are God's light descending to the earth. Alevi Zazas in Tunceli, Erzincan and Bingöl kiss the stones and rocks that the first rays of the sun touch and show respect to them. According to a common belief of the Zazas about solar eclipses, the reason for the solar eclipse is that dark spirits block the sun and prevent its rays from reaching the earth. Alevi Zazas believe that shooting arrows into the sky, beating drums or tin cans will scare away the dark spirits and save them from the eclipse. In order to save the moon from the eclipse, people try to make noise by beating drums and tin cans, just like in a solar eclipse, and shoot arrows into the sky to drive away the evil spirits that block the moon. This belief and practice are seen among Alevi Zazas in Erzincan, Tunceli and Bingöl.

Another Zoroastrian feature is the **"Jâr"**, the sacred trees, water springs, lakes, rocks, mountains and cairns that are important pilgrimage destinations. A Jâr (Çâr) is a sacred natural object with which people feel a mysterious affinity. It has arisen in nature and is not created by a human being. A tree must not be felled and burned, a rock must not be broken and a cairn should remain untouched. Jâr cults play an important role in the daily life of Alevis. Visitors do not pray at a Jâr, however, they have great respect for it and kiss it. Men, women and children are expected to dress in their finest clothing when they visit a Jâr. During a visit to a Jâr, an animal **"Kurban"** is sacrificed. It is slaughtered and a part of the meat is donated to the participants and villagers. The remainder is prepared for family celebrations where relatives and friends are invited. The sacrificial animal must meet certain age and quality standards, otherwise the animal is considered unacceptable to victims. Usually, it must be at least 2 years old, although a camel must be at least 5 years old. Poultry are not considered sacrificial animals. People should be reconciled to each other during this day, primarily with their parents then with their other family members and friends. Instead of sacrificing animals, other similar practices include donating a piece of **"Lokma"** bread cake to participants and villagers when visiting a Jâr. The Jâr cult also occurs with the Zaza-Goranî people in the eastern parts of the Zagros Mountains in north-western Iran.

In the Avesta, the sacred book of the Zoroastrianism, and the Persian epic **"Shahnameh",** it is said that Zoroaster planted a cypress in Kashmar in north-eastern Iran to honour his patron king Vishtaspa. According to some sources, the tree was already a place of pilgrimage during the life of the Zoroaster. The cypress in Kashmar was cut down on the orders of the Abbasid Caliph al-Mutavakkil in 861, as he believed the tree was a source of idolatry and fought against Islam. The tree was cut down and transported to Samarra in present-day Iraq. In the harsh region of Eastern Anatolia, many areas are treeless.

Instead of the sacred cypress tree, the Alevis choose to attend a rock, cairn, deciduous tree or water spring.

There is a distinction between religions, which live and exist within a people or state, and others whose goals are worldwide distribution and adherence among all peoples. With different emphases and different interpretations of reality, the boundary line can also be fixed in terms such as natural religion and prophetic religion Zoroastrianism.

For the duration of the Greek Invasion in the 3rd century BC, also during the times of the Roman Empire which ruled from the tum of eras until the 4th century AD, as well as during the rise of Christianity, Zoroastrianism earned the status of a primitive religion, due to its dualistic perception of the warm and its practice of fire worship. However, Zoroastrianism was vibrant and practiced as a foundation of belief, on all social levels. All of these Empires followed the politics of ruling by implementing their own government above the national structures. Thus, the national structures remained and the religion could be relative freely practiced under the new rulers. During the Persian reign Zoroastrianism became very strong and rejected the growth of Christianity. This changed with Islam, as idolatry was the worst sin called ***Shirk*** (bring partnership to Allah). Yet the followers of Zoroastrianism were fiercely persecuted and annihilated.

Many fragments of shamanism or "Tengrism" exist among the Zaza people. One of these is, when a relative is seriously ill, one takes a rope around one's neck and prays in the moonlight for God to cure the dying sick relative. In the presence of a Dede at an Ayin-î Cem gathering, the most common rhythmic music is with **"saz"** and **"deyiş"** singing to facilitate the transition to a trance state, during which the journey to the spirit world/dream world is performed. Wish trees decorated with ribbons mark a great deal of respect of the Alevis in the provinces of Tunceli, Bingöl and Elazığ.

EARLY SHIA ISLAM

Shia Islam is a branch of Islam that shares almost all essential features with other branches. Here are the same five pillars that frame the religious life, the same basic beliefs, the same belief in the finality of the Prophet Muhammad's revelation, the same Quran, the same monotheism and the same pious aspiration to let Islam – as an overarching divine order – permeate the whole of life. As in Sunni Islam, there is here the tension between close to the text and freer philosophical interpretations of the sources, the competition between the understandings of the legal élite and the popular expressions. Here are advocates for both traditional and modern orientations, here are groups with Sufi but also with anti-Sufi traits, and here are both Islamist and secular ideas about the place religion should occupy in society. Shiite Islam is thus not a homogenous orientation within Islam, but a broad, multifaceted and dynamic tradition that contains many different expressions.

Nor do Shia Muslims form an insignificant minority among the world's Muslims. Globally they may not be more than 15% but in the Middle East about 40% of the population is Shiite. They form the majority of the population in Iran, Iraq, Syria, Azerbaijan, Bahrain and Yemen and there are large Shia minorities in Saudi Arabia and Lebanon as well as in Pakistan and India. Smaller minorities can be found, for example, in Afghanistan and East Africa. The total number of Shiites in the world is probably around 250 million people. The fact that Shia have often been presented as a minority or as a special interest in the media and in academic research on Islam reflects a historical and political order in which Shia have often found themselves in a subordinate position in society.

It is difficult to talk about the core or essence of a religion. Considered as cultural giants, religions have no cores. They are changing accumulations of narratives and behaviours that mean different things to different practitioners and in different contexts. This also applies to Shiite Islam – but perhaps, if only for the sake of pedagogical clarity, certain critical features must be discussed.

For Shiite Islam, the closeness and loyalty to the family of the Prophet Muhammad is a particularly distinctive one. For many, Shia Muslims, the remembrance and veneration of the ***Ahl al-Bayt*** (Turkish **Ehlî-Beyt**, Goranî **Ahl al-Haqq**), which is the term used for the immediate family and descendants of the Prophet, is an important part of religious life. Stories about the family members are told, portraits of them adorn the walls of many and their death days are celebrated. Since many of the family members lived under persecution and oppression, the veneration of their memory has come to be about this. Other clear characteristics within Shia are therefore the grief and the fight against oppression.

There is an extensive literature within the Shiite tradition that is used to find support for their interpretation of the early history of Islam. The Quran is of course the most central source but beyond this there is the Shiite hadith literature (stories of what the Prophet Muhammad and, in Shiite Islam, the Shiite Imams did and said). These are mainly four major such works, among which the most famous is ***Kitab al-Kafi*** (The Sufficient Book). This was compiled by a certain Muhammed al-Kulayni (864-941). When it comes to interpreting one of the very first stories, ***Kitap al-Sulaym*** (Book of Sulaym) is also an important source. Sulaym ibn Qays (628-695) was a companion of Imam Ali and the four Shiite leaders who followed him. His book is a compilation of hadiths that clearly support the

Shiite understanding of developments after the death of the Prophet Muhammad. There is also a collection of books attributed to the Shiite leaders themselves. Special mention should be made here of ***Nahj-ul Balagha*** (The Way of Eloquence), a collection of sermons and letters attributed to Imam Ali ibn Abi Talib. The text was compiled by a certain Ali al-Radi (970-1015) but is considered within Shia to fall back on orally transmitted and authentic Ali quotations.

The Rise of Shia Islam

To grasp the origins of Shia Islam as a tradition and its focus on the Prophet's family, grief and oppression, one must turn to the Prophet Muhammad and his relative and close ally Imam Ali ibn Abi Talib.

Imam Ali was the son of Abu Talib who was the Prophet Muhammad's uncle and the one he had grown up with because he was orphaned at an early age. Imam Ali was thus not only the Prophet's cousin but also his foster brother. When Imam Ali was born, however, the Prophet Muhammad was already a grown man and in Shia it is usually believed that the fact that he was partly raised by the Prophet himself, made Imam Ali particularly suitable as a leader after the death of the Prophet Muhammad. Imam Ali was also one of the very first to believe in the Prophet Muhammad's message and become a Muslim. He was then just over 10 years old. Imam Ali fought by the Prophet's side in several battles and he became famous for his bravery on the battlefield. In addition, he married Prophet Muhammad and Khadija's daughter Fatima and together they had two sons Hasan and Husayn.

In Shiite tradition, one often talks about the unique community that existed between these five people – Muhammad, Fatima, Ali, Hasan and Husayn. They are called "People of the Cloak" **(Ahl al-**

Kisa) because, according to a hadith, the Prophet Muhammad once gathered them all under his cloak, prayed for them and called them the people of his house **(Ahl al-Bayt)**. It is this that, according to Shiite tradition, is referred to in the Quran when it says "God wants to free you, you who are closest to the Prophet, from all (earthly) dirt and make your purity perfect". Within the Shia, this is considered a holy and unique family who, because of their sinlessness and closeness to the Prophet, were destined to carry on Islam after his death. It is thus emphasised that the Quran and the Prophet's Sunna (Turkish **Sünnet**) are the documents of Islam, but at the same time believe that it is the ***Ahl al-Bayt*** (Turkish **Ehlî-Beyt**) who are the main sources for the interpretation and understanding of these. Hazrat Fatima is of particular importance here. As the Prophet's daughter, Imam Ali's wife, and supplicant of all Imams, she is the most important woman in Shiite historiography. She is described as a role model in terms of courage, uprightness and godliness and in Shia tradition it is said that she was killed by Umar in connection with the conflict after the death of the Prophet.

According to Shiite historiography, it was God's plan that Imam Ali would become the leader of Islam after the Prophet Muhammad. Within the Sunni tradition there is a different view on this and the question of who was the rightful successor of the Prophet is therefore the most debated issue in the dispute between Sunni and Shiite Islam. An event that Shiites often highlight in support of their interpretation took place at the oasis of Ghadir al-Khumm, located between Mecca and Medina. It is narrated in the hadith literature that Prophet Muhammad in the autumn of his age, after he had made the pilgrimage to Mecca for the last time in his life, on his way home to Medina he stopped at Ghadir al-Khumm to speak to his followers. Shaded by palm leaves held up above him, he asked them if he was not closer to them than they were to themselves. They then shouted back that this was the case. Then Prophet Muhammad invited Imam

Ali to stand next to him, lifted his hand and said: "Whoever has me as his ***Mawla***, his ***Mawla*** shall also be Imam Ali". According to some reports, the Prophet must have then added a short prayer: "O God, be the friend of his friend and the enemy of his enemy".

Shia and Sunni scholars agree that this must have occurred, but there is a difference of opinion on how the word ***Mawla*** is interpreted. According to in Shiite tradition the word should be understood as "Leader", while in Sunni tradition it is rather interpreted as "Friend" or possibly "spiritual leader". Shiites believe that the Prophet Muhammad at Ghadir al-Khumm conveyed to the Muslims what he had been told by God, namely that Imam Ali was the one who was destined to shoulder the leadership of the Muslims after his death. Shiites believe that this interpretation is strengthened by the Quranic verse that the Prophet Muhammad is said to have been revealed on the same occasion: "Announce, O Messenger, all that your Lord has revealed to you". It is therefore explicitly about the message that Imam Ali is the chosen one and not about the revelation in general. This interpretation is strengthened, according to Shiite understanding, by the next line "If you do not do that, you have not conveyed His message. God will protect you from the people. God does not guide the deniers of the truth!". Here it appears that the message, even though it comes at the end of the Prophet's life and after the Muslims have triumphed over the resistance from Mecca, is controversial. According to Shiite interpretation, this shows that it can hardly be about Islam in general, but precisely about the appointment of Imam Ali.

Sunni historiography therefore does not deny the incident at Ghadir al-Khumm or that the Prophet Muhammad would have expressed himself in this way. However, there is a different interpretation of what this means. Imam Ali, as the fourth rightly guided Caliph, is an elevated and important person also within Sunni Islam, and the Prophet Muhammad's recognition of him confirms that this is correct, but without thereby legitimising the entire Shiite interpretation. Within secular religious historical research, the interpretations of what the Prophet's intentions were also diverge. The most common is to interpret history in a way that is closer to the Sunni historiography, but there are also influential historians who align themselves close to the understanding of the Shia Muslims.

After the death of the Prophet Muhammad, history took a different development than that which God and the Prophet, according to Shiite historiography, had wanted. A distinctive feature of Shiite theology is the principle of ***Bada***, which means that God's plans can be changed and that it is ultimately up to us humans to implement them in history. Abu Bakr, another of the Prophet's allies, was chosen by a group of Muslims to be his deputy (Caliph) and the new leader of the Muslims instead of Imam Ali. When the election of Abu Bakr took place, Imam Ali was not present, he was in the Prophet's house to watch over his dead body and to prepare the burial. According to Shiite historiography, the election of Abu Bakr was thus unjust and against the will of the Prophet. Imam Ali also refused for 6 months to recognise Abu Bakr's leadership. According to the 9th century Persian historian Muhammad al-Tabari – whose writings form a very important source of the early history of Islam – Imam Ali stated that, he felt hurt by the rule of Abu Bakr and later his successor Umar over him and the Prophet's family, but that he forgave them for this abuse. However, this position reflects a Sunni rather than a Shiite understanding of the course of events.

Despite this, it is the schism between Imam Ali and the first three Caliphs that is the origin of the split between Sunni and Shiite Islam. The word Shia roughly means follower in Arabic and it refers to ***Shiat Ali***, Ali's followers or party, that is, those who, in the question of who would lead the Muslims after the Prophet, sided with and gave their oath of allegiance to Imam Ali. The conflict between the different orientations escalated after the third Caliph Uthman's reign. According to Muhammad al-Tabari's account, Imam Ali believed that

Uthman, unlike his predecessors, aroused the discontent of the people due to his poor leadership and that he was therefore assassinated, after which the people came to Imam Ali and asked him to shoulder the mantle of Caliph, which he reluctantly did.

Imam Ali's 5 years as Caliph became a turbulent time marked by internal strife between the various groups. Uthman's clan did not accept Imam Ali's leadership which led to war between the Muslims. In the Battle of Siffin in 657, Imam Ali's army fought against Muawiya. Muawiya was Uthman's second cousin and later became the first Ummayad Caliph. It was a bloody battle but no definitive victor could be crowned and the question of who was entitled to the title of Caliph therefore remained unresolved and disputed for the rest of Imam Ali's life. There are several levels to this conflict. The obvious reason, of course, is the dispute over who was the rightful leader after the Prophet Muhammad.

But during that conflict there were other levels – the warring parties belonged to different tribal groupings and they were also connected to different parts of the Prophet's immediate family, Imam Ali's wife Hazrat Fatima was the daughter of the Prophet Muhammad and his first wife Khadija, while Aisha, who sided with Uthman's family, against Imam Ali, was married to the Prophet after Khadija and also the daughter of Abu Bakr. Furthermore, there was a connection back to the very first years of the Prophet Muhammad's work. Parts of Uthman's family, including Muawiya, had in their youth been part of the group that persecuted the Prophet Muhammad in Mecca and that drove him to flight **(Hijrah)** in 622. A flight during which the young Imam Ali took the Prophet's place to confuse and face his persecutors in his stead. There was thus also an old personal enmity between the two Caliph pretenders.

The Imamate

In addition to the concrete political causes of conflict, a doctrinal and theological dimension has also emerged over time in the conflict between Sunni Muslims and Shia Muslims. The question of Imam Ali's right to leadership is, in Shiite tradition, ultimately about what is the essence of the religion that the Prophet Muhammad revealed to humanity. Herein also lies the theologically most essential difference between Sunni and Shiite Islam, namely the idea of the Imamate.

Islam's confession of faith, the ***Shahada***, which is called out in prayer in Muslim countries, after all, consists of the stanzas "I bear witness that there is no god but God", and "I bear witness that the Prophet Muhammad is God's messenger". These phrases summarise the main idea in Islam as a creed, the belief in the unity of God and the finality of the Prophet Muhammad's revelation. Within the Shiite tradition, an additional phrase is usually added to these two, namely "I bear witness that Imam Ali is God's chosen friend". The Arabic word translated here as "friend" is ***Wali***. This is a difficult to translate and ambiguous concept, related to the ***Mala*** discussed above. On the one hand, a ***Wali*** is someone who is particularly close to someone. In Sunni tradition, it has often been used for important Sufis to mark their closeness to God. But the word is also associated with the concept of Wilaya which means authority or leadership. When Shia Muslims say that they confess that Imam Ali is God's ***Wali***, they thus express partly that Imam Ali is particularly close to God and partly that he possesses a special divine authority.

These two ideas also express the main idea of the dogma of ***Imama***. Namely, according to Shiite tradition, the Prophet Muhammad left behind two sources of continued religious guidance. This is recited in the central Shiite "hadith of two heavy things". In this the Prophet says that he leaves behind two heavy things. The one is the recorded divine revelation in the Quran, and that others are ***Ahl al-Bayt***, his family. Sunni scholars interpret this hadith differently but within Shia it is seen as proof of the legitimacy of the idea of Imamate.

The Imamate idea means that the Quranic revelation cannot be understood without the guidance of religiously perfect leaders. As long as the Prophet was alive, his perfect example provided this guidance but after him, say the Shiites, God has not left mankind alone but continues to guide the people through a series of infallible leaders from the Prophet's own family (Ahl al-Bayt). The idea here is that the people around Prophet Muhammad were not ready to fully understand the revelation that he conveyed but that it had to continue to be expounded and explained by sinless leaders (Imams). Imam Ali, who was the first of these, also, according to Shiite sources, took over two esoteric texts named ***al-Jamia*** and ***al-Jafr*** from the Prophet. In these the perfect knowledge of the world, religion and the meaning of the revealed books were recorded. The task of Imam Ali and his followers therefore became to convey this knowledge to humanity at the rate that people are ready to receive it. His leadership will, therefore, be the beginning of a new era in the relationship between God and his creation. With the act of the Prophet Muhammad, the era of prophethood thus ends, while the era of the representative Imams begins. From a Shiite point of view, then, Imam Ali's loss of power was not just a political injustice, but a violation of God's plan for humanity.

The word ***Imam*** simply means "leader" in Arabic. It is used in many different ways in Islam. In Sunni Islam it usually refers to a person who leads the prayer. It can also be used as an honorific for particularly important teachers and scholars throughout history. Within Shia, however, the term has primarily come to refer to the unique collection of persons who constitute humanity's infallible leaders. Imam Ali was thus the first of these according to this tradition. According to Shia theology, God has appointed the Imams from the beginning of time, but they are identified in this life by the previous Imam naming his successor in a special statement.

The Tragedy of Karbala

The sources tell us that Imam Ali was murdered early one morning when he was about to pray in the mosque in the Iraqi city of Kufa. The assassin, Abd al-Rahman ibn Amr ibn Muljam al-Muradi, belonged to an extremist group known as the Kharijites (Turkish **Hariciler**). These had previously been loyal to Imam Ali in the conflict with Uthman's family, but abandoned him as they considered him to have made too great concessions to his enemies.

After the death of Imam Ali, his son Imam Hasan assumed the role of Imam of the Shiites. In the ongoing conflict over the Caliphate, he also became a pretender to his father's Caliphate and a competitor to Uthman's relative Muawiya. However, Imam Hasan soon chose to give up his claims and instead negotiate peace with Muawiya until a truce was reached. According to Shia historiography, Imam Hasan was still the rightful leader, but he made concessions on the political level to spare the Muslims unnecessary bloodshed.

When Imam Hasan died in 670, the leadership of the Shiites passed to his brother Imam Husayn. Now began an event in the history of Islam that has come to have great significance for the Shiite tradition. When Muawiya died in 680, the title of Caliph was taken over by his son Yazid ibn Muawiya. For Imam Husayn this meant that Imam Hasan's peace settlement with Muawiya no longer applied and that the question of who was the leader of the Muslims was once again open. Imam Husayn refused to swear an oath of allegiance to Yazid and thus the conflict flared up again. In 680 Imam Husayn received an invitation to travel to Kufa where since the time of Imam Ali there were many who were loyal to him. Imam Husayn accepted the invitation and set off on a journey through the desert. The entourage that travelled from Mecca consisted, apart from the Imam himself, of his immediate family – including his sister Zaynab (626-683) and his five children, as well as a number of companions.

Figure 33. The Imam Husayn Shrine – Karbala/Iraq

However, the entourage never reached Kufa. Instead, they were surrounded at Karbala, a place in the desert, by an Ummayad army sent out by Yazid. It is said that during the first days of the month of Muharrem they suffered severe hardships. Their carried water soon ran out and Imam Husayn's half-brother Abbas ibn Ali, who was sent out to fetch water, was brutally killed by Yazid's men.

On the morning of the tenth day of the month of Muharrem in the year 61 after the Hijrah (680) the battle began. According to Shiite sources, Imam Husayn's group consisted of 72 fighting men. Against these stood an Ummayad army with tens of thousands of soldiers. It was a bloody battle which, despite the courage and skill of Imam Husayn and his men, ended with all the fighting men and several of the non-combatants, including Imam Husayn's newborn son Abdullah, mercilessly killed, after which their bodies were desecrated and decapitated. Imam Husayn's body is today buried in Karbala, where his tomb is one of Shia Islam's most important pilgrimage sites. His head, however, was impaled on a spear and taken to Yazid in Damascus. The Umayyads had their headquarters in Damascus. Later, according to some sources, this relic was taken to Egypt and today there is a shrine built around it in Cairo. Other sources, however, believe that the head was taken to Karbala or remained in Damascus.

The martyrdom of Imam Husayn and his 72 companions in Karbala has become the Shia's worst historical reference point. The stories of the suffering endured by the Prophet's family have become a passion story through which not only the early history of Islam, but also human existence in general, is interpreted. Imam Husayn's martyrdom is understood as the act by which Islam was saved, his willingness to die against all odds to stand up against oppression and for truth and justice is seen as the prime example of what is true Islam. In the everyday and ritual life of Shia Muslims worldwide, the memory of the Karbala drama has also come to play an increasingly important place throughout history. The first 10 days of the month of Muharrem are remembered every year for the sacrifice of Imam Husayn and on ***the Day of Ashura*** it is customary for Shia Muslims to go in procession through the communities where they live to show their loyalty with the Prophet's grandson and to stand up against various forms of oppression that exists today.

Zaydis, Ismailis and Imamites

However, the succession of Shiite Imams did not die out with Imam Husayn. His son, best known as Zayn al Abedin (659-712), survived the Karbala massacre and became the fourth Imam. After him the succession continued from father to son. The Imams thus continued to lead the Shia Muslims but they lived under the strict watch of the Caliphs and none of them tried again to challenge political power in the way that Imam Husayn did in Karbala.

Shia Islam can thus be said to be the collective name for the Islamic traditions that believe that the rightful leadership after the Prophet goes further in a succession among the descendants of Imam Ali. Throughout history, however, disputes have arisen on several occasions about what this succession looks like. This has led to the emergence of a number of different Shiite groups.

One grouping is the so-called Zaydites, named after Zayn ibn Ali (695-740) whom they believed to be the rightful fifth the Imam. This group, doctrinally close to Sunni Islam, regards the first three Imams Ali, Hasan and Husayn, as infallible and chosen by God, but believes that the role of Imam should thereafter be understood as a political leadership role that can be borne by each Imam Ali and Hazrat Fatima-descendant who stands up against oppression. The focus is, thus more on the issue of justice struggle than on the idea of a divine knowledge that is managed in a certain lineage. Today, Zaydis are mainly found in Yemen.

Another grouping is the Ismailis. It is a designation for them various Shia groups who believe that a certain Ismail ibn Jafar (719-775) was the rightful seventh Imam. This orientation has played an important role in the history of Islam. In the 9th century, Ismaili leaders founded the Fatimid Caliphate centred in Egypt, and throughout history various Ismaili groups have played an important role in cultural and political development on and off. Today, Ismaili Islam is divided into a number of different orientations. The largest of these are the Nizarites, whose current Imam is Prince Karim Ağa Khan IV (born 1936). In recent years, this approach has acquired a liberal and modernist profile and has contributed strongly to cultural and scientific development in the Muslim world. Other groups that can be counted in the Ismaili family are Bohra Muslims and Druze. Most Ismailis live in South Asia or Central Asia, but large groups also reside in East Africa and in Europe.

By far the largest among the Shiite groups, however, are the Twelver, ***Ithna Ashariyya***, who are also called Imamites. Twelver Shia make up a large majority of the world's Shiites. According to the Imamites, there have been a total of Twelver Imams of whom the last one is still alive. The names of the Twelver are as follows:

1. Ali ibn Abu Talib, died 661
2. Hasan ibn Ali (son of Ali), died 669
3. Husayn ibn Ali (second son of Ali), died 680
4. Ali Zayn al-Abidin ibn Husayn (son of Husayn), died 713
5. Muhammad al-Baqir (son of Zayn), died 735
6. Jafar as-Sadiq (son of Muhammad), died 765
7. Musa al-Kazim (son of Jafar), died 799
8. Ali al-Rida (son of Musa), died 818
9. Muhammad at-Taqi al-Jawad (son of Ali), died 818
10. Ali an-Naqi al-Hadi (son of Muhammad), died 868
11. Hasan al-Askari (son of Ali), died 873
12. Muhammad al-Mahdi (son of Hasan), born 869, hidden 874

There is not space here to go into detail about all the different Imams and their life deeds. However, in addition to those already mentioned, there are two as, because of the meanings they have acquired in history, deserve to be especially be noticed.

Figure 34. The Twelver Imams

Jafar as-Sadiq and the Shiite School of Law

The first of these is Jafar as-Sadiq, the sixth Imam. He was a scientist and intellectual who lived at the very beginning of the period that is sometimes called the formative period of Islam, or the cultural golden age of Islam which often extends to the fall of the Abbasid Caliphate in 1258. Jafar as-Sadiq wrote books and taught astronomy, medicine, philosophy, chemistry, theology and Islamic law. He is recognised for his learning not only in Shia circles but also in Sunni Islam, and he is usually described as the founder of many ideas in Shia theology, including the aforementioned idea of the Imamate. The Shiite school of law **(Madhhab)** is also called the ***Jafarite*** and is named after him. Jafar as-Sadiq was also the teacher of the important Sunni jurists Abu Hanifa (699-767) and Malik ibn Anas (711-795), which is an example of the boundaries while what would later become Sunni and Shia Islam then was still very fluid just as the foundations of the Sunni and Shiite legal traditions are not necessarily very clear.

The Jafarite school of law is thus a fifth school of law alongside the four Sunni schools. In most matters these schools are very close to each other but there were some areas where the Jafarites differ from the Sunni schools. For example, in Jafari law it is accepted to combine the five daily prayers so that these are prayed three times during the day. Practising Shia Muslims therefore usually pray not at five, but at three times per day, before sunrise, at midday and after sunset. Another feature concerns the rules surrounding marriage. In Jafarite law, there is, alongside ordinary marriage, something called temporary marriage. This is used to give legitimacy to shorter or temporary sexual relations but can also be an opportunity to test a relationship without making it permanent through marriage. The conditions for these marriages differ significantly between the sexes, for example the bride must be unmarried while this is not a condition for the man.

Muhammad al-Mahdi and the Revelation

The twelfth Imam, Muhammad al-Mahdi, has a particularly important role in Shia Islam. Historically, the circumstances about his life are unclear, but in Shiite tradition it is said that the eleventh Imam Hasan al-Askari at his death in 873 had a young son named Muhammad. Only five years old, this Imam took over the role of Imam from his father, but for various reasons he chose to remain hidden and to communicate only with his followers through four selected mouth-pieces. It is said that in 941, when the last of the four spokesmen died, the Imam was included in what is known as the great concealment. From this day on, the Imam is believed to live on in the world as a living leader with the same authority and unique role, but hidden from both his enemies and his followers.

In Shiite tradition, the belief in the hidden Imam Muhammad al-Mahdi has merged with the messianic tradition found in all Abrahamic religions. Hasan al-Askari's son here becomes Muhammad al-Mahdi, the righteous saviour who will one day return to set everything right and who is also expected in Sunni Islam. Shia Islam teaches that he is ***Sahip al-Zaman*** (the Lord of Time) and ***al-Hujja*** (the Evidence), that he has been given extended life by God, and that when he returns, along with Jesus Christ, he will fight against the Antichrist **(al-Dajjal)** and herald the Day of Judgement. The proof of his authenticity when he appears will be that he has in his possession the scriptures ***al-Jamia*** and ***al-Jafr*** which have been passed down in the Imam succession, he will also fulfil a series of prophecies that previous Imams and the Prophet said about his return.

The notions of Muhammad al-Mahdi are understood in different ways within Shia Islam. There is certainly consensus about his concrete existence, but the wider meaning of this is understood on different levels. There are theological expositions in which his return is interpreted as an expression of man's potential to achieve the state of wideness and peace that we long for and there are esoteric traditions that speak of him as existing on another inner plane of consciousness. On a popular level, a plethora of more or less sanctioned stories also exist about how Muhammad al-Mahdi has appeared to people in dreams or on particularly critical occasions. It is also a widespread belief that certain particularly exalted and respected jurists are in direct contact with him.

Shia Islam after the Imams

Although the origins of the schism between Shia and Sunni Islam lie right at the beginning of Islamic history, it is a simplistic misconception to see today's division between these two groups as a straight continuity from the conflicts of the 6th century. Instead, the emergence of today's situation should be seen as the result of a political polemic between different groups that developed over barely a millennium and a half. In fact, the line between Shia and Sunni Islam for many centuries has been, and today partly still is, very vague. There are many groups that are in the borderland between these greats, in addition to the Zaydis, the Alevis in Turkey and the Alawites (Nusayris) in Syria, as well as the many Sunni Sufi Orders whose tradent chain (chain of transmitters of hadith) goes through the Shiite Imams can be mentioned as examples. To understand how today's Shiite groups were formed, one must therefore also study the later political history of Islam.

During the first millennium of Islamic history, a number of different Shiite kingdoms were formed. Here the Zaydi Idrisi dynasty in present-day Morocco (788-974), the Ismaili Fatimid Caliphate centred in Egypt (788-1171), the Persian Buwayhids (934-1062) and their allied Arab neighbouring dynasty the Uqailids (990-1096) are noticeable. Some rulers of the Mongol-founded Ilkhanate (1256-1335), which covered large parts of the Near East, were Shia influenced, as were the regents of the Bahmani Sultanate in India (1347-1527) and, eventually, the kings of the Turkish Safavid Empire in Iran (1501-1736).

It is in the political history of these kingdoms that the conditions for much of the polarisation between Shiites and Sunnis that we can witness today were created. By adopting a Sunni or Shiite identity, rulers have been able to legitimise their struggle against other Muslim dynasties. Many of today's Shia Muslims in Iraq, for example, are descendants of groups that, because of their dissatisfaction with Sunni supremacy, converted to Shia in the 17th and 18th centuries. So almost a millennium after the original conflict between Imam Ali and Caliph Abu Bakr.

THE OTTOMANS (1299-1923)

With the fall of the Baghdad Caliphate, the Arab-Persian areas where Islam had grown lost even more of their influence. Even earlier, the Turkish Seljuks had been successful, and with the establishment and expansion of the Ottoman Empire in the 14th century, Turkish groups came to play an increasingly important role in the spread of Islam to the west. Turkish also gradually became one of the dominant languages of the Muslim world alongside Arabic and Persian. Through the conquest and transformation of Constantinople to Istanbul in 1453, the Ottomans definitively took over Byzantium's dominant role in the Eastern Mediterranean and were able, through further expansions, to connect parts of the Middle East, North Africa and South Eastern Europe. This strategic position was to be maintained by the Ottomans for centuries to come. Relations with Europe were rarely uncomplicated, but contacts were undeniably there through trade, diplomacy and war.

Regarding the Zaza people and the Alevi Kurds, the Ottoman Empire created a form of government that functioned as a system of protection against external enemies such as the Persians, Russians or other powers. Internal politics focused on a tribute-based tax system and general obedience, but allowed for self-government. The Zaza people were independent and self-sufficient people, as they governed their small-scale society. These were the days when the Ottoman Empire was popular as an interfaith, intercultural and liberal regime. This naive and blue-eyed picture of the Ottoman Empire does not reflect the truth.

From those days on, many attacks on non-Islamic groups of people have been reported. From an anthropological point of view, it was jealousy driven by irrational xenophobia that determined the relationship of the ruling system with the ***Millets***. The fear of losing the welfare and well-being of non-Islamic groups was often due to the fact that it was easy to plunder unarmed people because they were second-class citizens and no criminal prosecution was expected.

From Nomads to State Builders

Under the threat of the advancing Mongols, many Central Asian Turkic tribes had migrated westward, reaching Anatolia in the 12th century. These nomads were made up of many small units that were mutually fragmented. But in 1301, Osman Gazi (1258-1326) took power over the Sultanate that the Seljuks had founded in Asia Minor and thus became the ruler of a large contiguous area. He gave his name to the dynasty that ruled the kingdom for 22 generations for over six centuries until its final collapse in 1923. A more legendary historiography usually portrays Osman Gazi as Sufi-oriented and refers to a dream he had of a large tree growing out of his umbilical. A Sufi Shaykh is said to have interpreted the dream for Osman Gazi and held that it was a prediction that his kingdom would grow and with it, Islam would continue to spread.

This state-founding myth fits with the ideal of being a great military leader **(Gazi)** cultivated throughout Ottoman history. The Turks were skilled warriors, swift on their horses and efficient with their bows. The military organisation was from the beginning a unifying factor among those who would go down in history as the Ottomans – and remained so. They could not claim kinship with the Prophet Muhammad like other Muslim dynasties but their legitimacy was based on political success. Under Osman Gazi, expansion continued westward and northward, towards the Byzantine Empire weakened by the Latin depredations of 1204 in conjunction with the Fourth Crusade.

In 1326, the son Orhan Gazi captured the city of Bursa, which became the first capital of the Ottomans, and he was also the first among the Turks to call himself ***Sultan***, which is an Arabic word for ruler. With this, the previously nomadic Ottomans took the definitive step towards urbanisation and state formation. Their kingdom was characterised from the beginning by an efficient administration with networks of schools, minting, tax collection and registration of subjects. The latter is today an invaluable source material for creating a picture of Ottoman everyday life – also out in the provinces.

During the 1350s, the Ottomans crossed the Dardanelles and entered Europe. The various ethnic and religious groups that existed in the conquered areas could largely continue to live as before as long as loyalty to the Sultan and the state was unbroken. In the early Ottoman Empire, Christians were in the majority and they continued to form a significant minority until the last days of the empire. The city of Adrianople (Edirne) in Eastern Thrace was the second capital of the Ottoman Empire from 1369 to 1453, before Constantinople became its capital.

The so-called ***Devşirme*** system was a forced military recruitment where Christian boys, mainly from the Balkans, were forcibly taken from their families and converted to Islam. This was considered a kind of tax compensation for Christians not doing military service, but ceased during the late 17th century. The boys were sent to school to learn Ottoman Turkish and many times became very well educated. Those who attended the best schools could continue with language studies and also immerse themselves in Islamic studies. Many of these young men were given the opportunity for a career beyond the military in the Janissary troops. Several came to go very far and stand close to the Sultan and the court, some even ended up as grand viziers.

Constantinople the Capital of the Ottomans

During the 15th century, Ottoman expansion continued in the Balkans and in 1453 Constantinople was taken. The Ottomans thus had full control over the region. Already during the late 660s, the Muslim armies had tried to capture Constantinople with its strategic location between the Black Sea and the Mediterranean. The Prophet Muhammad's standard-bearer, called Eyüp Sultan in Turkish and Ayyub al-Ansari in Arabic, according to Islamic historiography, must have fallen outside the city walls in connection with the siege in 674. His grave must have been revealed through a dream upon the arrival of the Ottomans. Eyüp's mausoleum **(Eyüp Sultan Türbesi)** is still today a popular pilgrimage destination just west of the Golden Horn and is counted after Mecca, Medina and Jerusalem as Islam's holiest places. The mosque complex also contains relics from the Prophet Muhammad, including his footprints. With such a reference to early Muslim history, the Ottomans gained a direct connection to the Prophet Muhammad through the new capital, and here also new Sultans were dubbed with Osman Gazi's sword.

The entry into Constantinople meant the definitive end of the more than the thousand-year-old Christian Byzantine Empire. There are many legends about the conquest and how Constantinople became the seat of a Muslim Empire that would last until the aftermath of the First World War. Sultan Mehmet II's (1432-1481) – called Fatih Sultan Mehmet – use of gunpowder and cannon made the Ottomans militarily superior even though the city's walls held a month-long siege. It was a bloody conquest and of those who survived many were sent into exile, but Sultan Mehmet II was quick to invite new residents for the reconstruction and development of the city's economy. Among them, Jews who were expelled from Spain in 1492 and who could contribute to strengthening trade contacts were welcomed. In all Ottoman cities of importance there were therefore significant Jewish groups. To symbolically mark the conquest, Sultan Mehmet II rode directly to Hagia Sophia (Turkish **Aya Sofya**), then the largest church in Christendom, and performed his prayer there, thus the church was considered converted into a mosque. Architecturally, however, the magnificent Hagia Sophia came to influence the design of the classical Ottoman mosques with its large domes. After Constantinople, expansion continued fairly immediately to the north and west with conquests of present-day Romania, Albania and Bosnia and it was a clear strategy on the part of Sultan Mehmet II to conquer surrounding Christian lands.

Many of Constantinople's churches were converted into mosques and new mosques were built to mark the Sultan's power and the city's importance in the empire. The transformation of the urban landscape can be exemplified by the construction of the mosque commemorating the city's conquest, the Fatih Mosque, the "Conqueror's Mosque". It was built on top of a destroyed church and with the ambition that its dome would be bigger than that of Hagia Sophia. In addition to being a significant memorial in the cityscape, the Fatih Mosque is a magnificent example of all the functions an Ottoman Mosque complex **(Külliye)** could accommodate in addition to being a prayer room in connection with the obligatory prayers of the day and, not least, for the Friday prayer with its sermon. In a ***Külliye*** there could be schools of different levels (**Mektep** for the smaller children and **Medrese** for more qualified teaching), library, ***Hamam*** (bathhouse), hospital, kitchen to feed the poor **(Imarethane)**. The activities around the great mosques provided for both religious and material needs. Within the mosque complex there were also often mausoleums **(Türbe)** for

Sultans, military leaders **(Paşa)** and pious men. The mausoleums were attractive local pilgrimage sites and the honoured dead were believed to spread blessing and healing power. So, for example, Sultan Mehmet II rests next to the mosque that was built in his memory.

Sultan Mehmet II was also the Sultan who asserted himself as the leader of the Ottoman Caliphate. Islam had a special position in the Ottoman Empire. It was the official religion the Sultan had to be Muslim (although several of them married Christian women) and ***Şeriat*** (Sharia) was the basic law. Starting with Sultan Mehmet II, the Ottomans acted as the leaders and representatives of the Sunni Muslim world and they were unchallenged in this position for several centuries.

The successor, Yavuz Sultan Selim (1465-1520), who is sometimes nicknamed "The Cruel One" (Yavuz), was the one who violated the earlier theological principle of not making war on other Muslims. In 1517, Yavuz Sultan Selim defeated the Mamluk Sultanate in Cairo and thus Egypt, Syria, Jerusalem, and Mecca and Medina (which remained under Ottoman rule until 1916) came under the Ottoman Empire. The Sultan thereby became the "protector of the two holy cities" – a situation not always appreciated in the Arab world.

The Ottoman Golden Age

During the 16th century, military power was consolidated and the Ottomans also began to become a cultural power. The very well educated, Sultan Süleyman the Great's long reign from 1520 until his death in 1566 is in many ways the pinnacle of the Ottoman Empire. After, as a young Sultan, he proved himself capable as ***Gazi*** with new military conquests, a general body of law **(Kanun)** was codified and spread throughout the empire. Sultan Süleyman is therefore known as ***Kanunî*** "The Lawgiver", in Turkish.

Sultan Süleyman favoured the arts and commissioned several of the finest buildings of the period, including the Süleymaniye Mosque in Constantinople, perhaps the most magnificent Ottoman building project with its six minarets. The foremost of the architects, Mimar Sinan (1489-1588), was probably born a Christian and was trained as an engineer through the ***Devşirme*** system. In a unique way, he combined beauty with technical genius and left his mark on monuments around the empire.

Islamic Hierarchy and Religious Minorities

The central Ottoman rule in Istanbul is sometimes referred to as the High Gate **(Babıâli)** and refers concretely also to the gate that foreign diplomats entered to get to various ministries and offices. The formal reception of foreign ambassadors began to develop into grand ceremonies during the time of Sultan Süleyman the Great when gifts were given and received. The long-standing diplomatic relations with various European states are a reminder of the frontier position of the Ottomans, not only in the east-west but also in the north-south directions. The splendour of the Sultan's court is a recurring theme in reports and travelogues.

The Ottoman Empire was fundamentally an Islamic state in that the ***Şeriat*** (Sharia) was the supreme law, and decrees and regulations could not go against the ***Ulama's*** interpretations of the ***Şeriat***. The Sultan was also the supreme patron of Islam and considered himself the leader of the Sunni world. The Ottoman legal tradition followed it the Hanafi school of law where "the principle of seeking to find the good or appropriate", along with analogies and con textualizations, is indicative. This made it easier to apply the ***Şeriat*** to changing circumstances in the expanding empire where different legal systems and theological traditions of interpretation were at least as significant such as the differences between Muslims and non-Muslims.

The Islamic administration was closely linked to the Sultan's court and the state. The central Sunni Muslim authority had the title of ***Şeyhülislam*** (Shaykh al-Islam) and constituted the pinnacle of the scholarly hierarchy of ***Ulama*** (Turkish **Ulema**) and had the authority to educate pronouncements based on the ***Şeriat*** that applied throughout the empire. ***Şeyhülislam*** would judge whether acts, decisions and decrees followed the ***Şeriat*** and he was the third in status after the Sultan and the Grand Vizier **(Sadrazam)**, but not a member of the Supreme Council **(Divan)**. Rather, his power was implemented by the network of ***Ulama*** that existed around the provinces. ***İlmiye*** was the central Islamic authority that would safeguard the spread of Islam and that given ***Fatwas*** were followed. This institution provided the basis for a certain theological consensus for the ***Ulama*** operating in the provinces and also constituted a structure within which ***Ulama*** and judges could make careers. In 1924, the Republic of Turkey abolished the ***Şeyhülislam*** (Shaykh al-Islam) office and replaced it with the Directorate of Religious Affairs **(Diyanet)**. Turkey thus did not separate Islam from the state, but shifted the model for controlling the public Muslim practice of religion.

The Ottoman Empire was multi-religious and multicultural from its beginning and eventually developed an administration to govern the non-Muslim groups. Every religious minority counted as one nation **(Millet)** which to a large extent had its own jurisdiction, at least in civil matters. The minorities had to pay a special tax **(Jizya)** which in return guaranteed them freedom to practise their religion. The "People of the Book" **(Ahl al-Kitab)** were subject to patronage and a person under patronage had at once limited rights in relation to Muslims, but protection within the civil code of his ***Millet***. In the

Ottoman cities it was common for the religious groups to live in their own quarters with a local authority and with definite rules for the interaction between different groups. Periodically, specific rules also applied to which clothes, colours and headdresses the various groups were forced or forbidden to wear. The ***Millet*** system gave the local ethno-religious groups some autonomy and power to the local leaders, but they were on a superior level subordinated to the Ottoman state administration.

Both ***Mektep*** and ***Medrese*** were many times connected to the ***Külliye*** complexes around the great mosques and were run by foundations **(Vakıf)**. Such foundations were absolutely crucial to both the charity and education that could be offered. In ***Mektep***, the local Imam could teach elementary Quranic recitation, how to correctly perform the prayer from ablution and intention to the correct movements and words. Girls could also attend the ***Mektep*** and were then instructed by a woman who might well be trained to recite the Quran by heart, but the home was otherwise the primary arena for women's instruction. However, they were not only taught Quranic recitation or basic knowledge of the content of other canonical literature. The repertoire used by women included poetry, genres of prayer other than the five obligatory and pious stories of the day, as well as sermons and expositions of the Quran.

The Sufi Orders were also important institutions of learning that were instrumental in the spread of Islam after the Ottoman conquests. Being transnational networks, they mediated the religious influences between different areas and local religious practices were often included in their rituals. Many different Sufi Orders were active within the Ottoman Empire, but two of them deserve special mention. The Mevlevî Order with the "dancing dervishes" was established in Konya in the 13th century. Their whirling in circles has, not quite rightly, become the image of Sufi piety in general. This form of prayer **(Sema)** is associated with the poet Mevlana Rumi (1207-1273) and is aesthetically refined in its verbal, bodily and spatial expressions. The Bektaşi Order, with roots dating back to pre-Islamic times, on the other hand, has had a more popular character and wide spread among the Alevi groups of Anatolia and in Albania, Kosovo, Macedonia and Bulgaria. Their theology is clearly influenced by Shia Islam, this and their rituals with both pre-Islamic and Christian influences were not appreciated by the ***Ulama***.

Figure 35. Ottoman Empire 1792

THE SAFAVIDS (1501-1736)

In the early 16th century, a coalition of a number of Turkmen succeeded tribes found a kingdom that extended over large parts of the ancient the Persian Empire that existed before the Arab invasion of the 650s. The Safavid Empire (1501-1736) included parts of the Caucasus, Eastern Anatolia, Persia and most of present-day Iraq and Afghanistan. The origins of the Safavids went back to a Sunni Sufi Order which was founded by the patriarch Safi al-Din (1252-1334) in the city of Ardabil in present-day Iran in the 13th century.

The Safavids proclaimed Shiite Islam as the state religion of their kingdom, and thus ended up as a wedge between the Ottoman Empire in the west and the Mongol-Uzbek rulers in the east, both of which were Sunni. Safavid Persia was now not only a military rival but also a religious competitor to the Ottomans who claimed to shoulder the mantle of the Prophet and to be the leader of all Muslims. The Ottomans attacked the Safavids on three occasions and took over areas in the Caucasus, Anatolia, present-day Iraq and north western Persia. However, the Ottomans did not achieve the feat of completely defeating their Shiite rival in the east.

The fact that the Safavids proclaimed Shiite Islam as the state religion was partly due to the fact that the Sufi Order that ruled over the **Kızılbaş** tribes in Eastern Anatolia had already contributed to the Shia in the 15th century as a result of living under it rule and patronage of the Shiite dynasty of Kara Koyunlu (1375-1468). The Safavids also invited Shia jurists from Jabal Amil (the region between present-day Syria and Lebanon), from the north eastern Hejaz (present-day Saudi Arabia), from the Two Rivers (present-day Iraq where several Shia Imams were buried) and other areas to take an active part in the administration of the kingdom. An important reason for this is believed to have been that the Safavids needed a religious legitimacy in the fight against the Ottomans and the Turko-Mongol Uzbeks.

It was not only the Safavid rulers who needed the support of the Shiite scholars. The dependence was mutual. The expansion of the Ottomans and anti-Shia policies had led to recurring killings and looting raids against Shiites. Many times, these attacks had been blessed by Sunni jurists. The plight of the Shiites made the invitations of the Safavids appear irresistible to the Shiite scholars. Under Safavid rule, they would not only be protected from the terror of the Ottomans and Sunnis, it was also a golden opportunity to nurture the Shiite minority, develop Shiite theology and also participate in the administration of a Shiite kingdom, which had never happened before.

The Corresponding Form of Governance

The participation of the Shiite jurists in the governance of the Safavid Empire laid the foundation for an administrative system that could be called a parallel form of governance. The parallel form of governance required that the legal scholars build up an organisation that could fulfil the tasks they were assigned. It was about various religious, legal, economic and social commitments, but also to some extent political such.

The Shiite jurists were soon faced with the question of how to relate to the Safavid kings **(Shah)**, such as Shah Ismail I (1487-1524) and Shah Tahmasp I (1514-1576) who had taken the initiative to invite the jurists to the kingdom. Based on Shiite political doctrine, all worldly rulers during the absence of the hidden Imam Muhammad al-Mahdi were considered illegitimate. The experience of the time of the four Imamite Shiite Buyids and the religiously indifferent Ilkhanid Mongol descendants had meant that Shiite jurists had modified their view on the matter and were not so categorical in relation to the worldly rulers. For example, they distinguished between righteous and evil rulers and opened the door to cooperation with the so-called illegitimate but righteous rulers. The Safavid kings followed the line of Shah Ismail I, the founder of the Safavid monarchy, and collaborated with the Shiite jurists.

However, there were those who did not want to give the jurists too much authority. Shah Tahmasp I, who ascended the throne at the age of 10 after his father Shah Ismail I and who, at least initially, was under the strong influence of the Kızılbaş forces (Alevi forces) and the Grand Vizier, assured in a letter to the Shiite jurist Noor al-Din Ali ibn Husayn Abdul Ali Karaki (1464-1533) that he regarded him as the representative of the Twelfth Shiite Imam and himself as the representative of Abdul Ali Karaki. Anyone who opposed this would be considered a heretic and severely punished, the king writes in his letter. For Shah Tahmasp I, the honours given to the top Shiite jurists were a way of standing up to the Kızılbaş forces and the Grand Vizier. The opposite of this was Shah Abbas I (1571-1629), who presented himself as a deeply religious Shia Muslim and who was very powerful. Shah Abbas I already circumvented the influence of the military and the Grand Vizier over the royal power at the time of his accession to the throne. He did not give the jurists much latitude and used their services selectively on the occasions that served his various political or military moves.

The situation of the jurists had been strengthened by the Safavid kings proclaimed Twelver Shia as the state religion and thereby pledged to defend Shia Islam against its Sunni enemies. These kings also showed great respect and reverence for the Shiite jurists and invited them to participate in the administration of the Shiite kingdom. Better conditions could not be wished for. But at the same time, the jurists wrestled with the dilemma of justifying their collaboration with Safavid kings who were by definition considered illegitimate. Another problem was that the jurists would legitimise royalty who did not care about the moral codes of Shia Islam in their private life. It was no secret that several of them met an early death due to alcohol abuse and overconsumption of narcotics, such as opium.

The Question of Legitimacy

The question that preoccupied a majority of legal scholars during the roughly 200 years rule of the Safavids was the question of legitimacy. That is, the question of on what grounds one would legitimise the Safavid kings. There was no consensus among the jurists who had agreed to participate in the parallel form of governance. There were those who chose to solve the problem from a strictly religious perspective. An example is the jurist Molla Muhammad Baqer Majlesi (1616-1699). During the coronation of one of the Safavid kings, he chose in his sermon to lay the foundation of legitimacy on the Safavid kings' alleged kinship with the Shiite Imams and ultimately with the Prophet himself. According to Muhammad Baqer Majlesi, this basis of legitimacy entailed important obligations for the Safavid kings. These would follow in the footsteps of the Shiite Imams and the Prophet and live a pure doctrinal life and be righteous. During his sermon in connection with the coronation of the Safavid king Shah Sultan Husayn (1668-1726) in 1694, he emphasised that God had chosen to give the kingship to this king, like other Safavid kings, so that the Shia Muslim subjects would live in security in the Safavid kingdom.

Muhammad Baqer Majlesi's legitimisation of the Safavids' position of power can be seen as a pragmatic solution where he tried to adapt the basic Shiite political tradition of ideas to the prevailing conditions. At the same time, it can be seen as a very clever solution to the question of the establishment of a worldly power that can receive the blessing of the jurists during the absence of the hidden Imam Muhammad al-Mahdi. In his sermon, he charged the king to live up to a series of criteria rooted in the Shiite religious tradition. However, he did not go so far as to elevate the Safavid king to the level of the hidden Imam Muhammad al-Mahdi or to his deputy. The main reason for this was that the Safavid kings, for all their leanings towards Shia Islam and despite their alleged kinship with the Prophet's descendants, could not be classed as sinless. This was reserved only for the Twelver Shiite Imams. However, Muhammad Baqer Majlesi's support for the Safavids was based on his belief that the Safavid era was a prelude to the return of the hidden Imam Muhammad al-Mahdi.

Another jurist, Baha al-Din Muhammad ibn Husayn al-Amin (known as Shaykh Bahai, 1547-1621), who worked at the court of Shah Abbas I solved this problem in a different way. Shaykh Bahai distinguished between personal faith and morality, or what he terms private justice, and public justice. Shaykh Bahai could conceivably turn a blind eye to the perversity of the kings in regard to personal faith and the anything but pure doctrinal life of court life as long as the kings in their public engagements followed the principle of justice in relation to their subjects and continued to defend Shiite Islam.

The Safavid kings thus had to content themselves with being presented as quite gracious rulers who could thus receive the blessing of the jurists. However, the blessing was conditional and applied only as long as the Safavid kings responded to Shia Islam. That the jurists did not stop the Safavid kings the level of the Shiite Imams meant that obedience to them was not equated with obedience to God and that crimes against them could not be seen as crimes against God. There were also Shiite jurists who felt that a legitimisation and blessing of Safavid rule could not be reconciled with basic Shiite political doctrine. The solution was to look at the leadership issue from a perspective where the focus was on safeguarding justice in society.

The scholars elaborated a series of qualities that a righteous king in a Shiite kingdom must have. This includes qualities such as uprightness, to follow the tradition of the hidden Imam Muhammad al-Mahdi, protect Shia Muslims from dangers and those infidels, cherish Shia and so on. By enumerating the qualities of the Shiite king and the demands placed on him, the jurists are freed from the requirement to tackle this task themselves.

Duties and Institutions

The Safavid kings' knowledge of Shiite Islam was extremely deficient, and in some cases completely non-existent. Their solution to this problem was to leave all religious matters to the jurists. However, they were keen to secure their power over the religious institutions. They were the ones who would appoint the highest religious leaders and they appointed them from among the high-ranking jurists who were most loyal to the royal house. This in turn could create divisive disputes between parts of the religious congregation and the royal house.

Despite the enmity with the Ottomans, the Safavid kings not infrequently looked towards their rivals in the west and copied them in many respects. Like the Ottomans, the Safavids created various religious institutions with specific tasks in legal or religious matters. By appointing the heads of these institutions, the Safavid kings left their mark on the involvement of the religious leaders in society.

One of these institutions in which the Safavid kings interfered was the appointment of the one among the Shiite jurists who would represent the religious institution of the royal house. This, who during the Safavid era was called ***Molla Bashi***, was the head of all Shiite jurists and had an administrative and representative function. At the audiences of the Safavid kings, ***Molla Bashi*** was the religious authority closest to the king. His task was to pray for the king's mercy on those who had fallen into disfavour, to pray for the innocent to be pardoned and to speak out on religious matters.

Another and more administrative position among the jurists were that office that went by the designation ***Sadr***, which was a state-appointed religious leader. His duties included appointing religious judges, responsible for religious fiefs, appointing prayer leaders, leaders of religious schools and mosques, and various other religious offices. By entrusting such religious matters to an official religious leader appointed by the Safavid court, the king was able to influence the development and supervision of the various religious affairs. At first it was the highest ranking in the Shiite religious hierarchy who were appointed to this office, but as administrative duties began to take over, this position became one among many other court positions that the highest religious leaders were reluctant to accept then the service was no longer seen as religiously significant. This was because the inclination of the jurists to take care of purely administrative posts was not as great as to hold a leading role in the religious, judicial or secular fields.

The office of ***Sadr*** fulfilled yet another important function dealing with the spread of Shia Islam. Proclamation of Shia Islam by the Safavids state religion took place in a kingdom largely populated by Sunni Muslims and ***Sadr*** thus had the task of ensuring that the public in the Safavid kingdom would convert to Shia Islam.

Another office of great importance during the Safavid era was ***Shaykh al-Islam*** (Turkish **Şeyhülislam**), the foremost of the jurists, the representative of the hidden Imam Muhammad al-Mahdi, whose obedience is incumbent on all Shia Muslims, whose duties included the coronation of the king. At the same time, it was the king who appointed who would receive the title of ***Shaykh al-Islam***.

The Fall of the Kingdom

In 1722, Shah Husayn (1668-1726) was defeated by Afghan tribal warriors after the Safavid capital Isfahan was besieged for eight months. Fourteen years later, in 1736, the Safavid Empire fell and thus the situation of the Shiite jurists changed dramatically. The fall of the Safavids meant not only the overthrow of a Shiite dynasty and the loss of the support of the Shiite jurists from the royal power. The Shia jurists were also subjected to severe reprisals for blessing the Safavid kingdom and for their complicity in the Safavid policy of consolidating Shia Islam in their kingdom. The new rulers saw the Shiite jurists as a religious and political obstacle that had to be removed. The fall of the Safavids was accompanied by political unrest, war, famine, disease, and part of the Shiite population being abducted and sold as slaves. The population of the former Safavid kingdom was decimated by about a tenth. The Afghan tribal warriors who dealt the final blow to the Safavids destroyed and vandalised great cultural, artistic and scientific treasures located in Isfahan, Iran.

The Safavids left their mark on the further development of what remained of their kingdom. Persia remained Shiite and the Shiite jurists and their institutions survived despite all the reprisals after the fall of the Safavids and came to play an important role in, among other things, the social and political development of the country.

CUSTOMS AND CULTURE

The great scene of the Neolithic agricultural revolution was that of Anatolia. From the transition between the Mesolithic and the Neolithic periods (about 10000-9000 years BC) up to about 5000 BC, single grain wheat was spread across a region from Anatolia to Southern and Central Europe, westwards towards the Atlantic coast. Since around 11,500 years ago, the eight Neolithic founder crops, einkorn wheat, rye, barley, peas, lentils, vetches, chickpeas and flax, were cultivated in the Near East. Sheep and goats were domesticated in Mesopotamia between 13,000 and 11,000 years ago. Cattle were domesticated from the wild aurochs in the regions of Anatolia and Pakistan approximately 10,500 years ago.

In Mesopotamia, according to recent observations, river water had already begun to be channelled in the upper reaches of the Euphrates and Tigris and their tributaries 8,000 years ago. This was done on a scale that was well within the reach of the village community or even of the individual family. The gradual conquest of the river valleys is not the only agricultural contribution to the emergence of high cultures. In addition, a number of new crops were added, which increased both the supply opportunities and the need for trade.

Basically, the Zaza people have kept to agriculture, animal husbandry and handicraft. The Zazas are permanent settlers and do not fit into the template of nomadic people, in contrast to the Kurds, who are nomadic people and migrate with their grazing animals from mountain to mountain. Zaza villagers had and still have their fields and small meadows scattered at all points of the compass. However, it was a highly barren and miserable farmland with very poor soil. There was a one-to-twelve ratio of sow to harvest. As the Zazas cultivated grain, they made canals or ditches which they filled with water from the rivers near them. The fields were ploughed with a **"kara saban"** (primitive wooden plough) driven by a pair of oxen. The villagers cultivated rye, wheat, barley, millet, oats, lentils, chickpeas, common vetches and so on. Beyond this, on the low plains the farmers grew leguminous plants, beets, watermelons, tomatoes, grapes, mulberries, apricots, peaches and so on. The seed corns were treated with copper sulphate (highly toxic) before sowing. Milk production occurred only in the summers as animal udders dried up during the winter season. Oxen were kept as draft animals for the ploughs and carts and for producing meat. Cows, sheep and goats were used to provide milk, meat, fat, wool and hair. The Zazas of Eastern Anatolia use ***tezek*** (dried dung of cattle and sheep) as fuel for heating and cooking.

Figure 36. Zazaic Burial Ground – Pülümür/Tunceli

Production of Iron

The wrought iron became known to the Zaza people around 2500 BC. The people were mainly farmers and craftsmen, especially the blacksmiths. The blacksmiths carried the traditional ironwork further. Iron smelting was winter work for the farmers when the agriculture rested. The Zaza settlements that became known for their iron forging were Kığı, Hınıs, Varto, Palu and Mazgirt. Eastern Anatolia during the period of the Hittite empire was an early centre for the development of ironmaking technology. For example, iron axes, iron pickaxes, iron arrow tips, iron shears, iron sickles, iron scythes, iron knives and iron ploughshares were manufactured. Kığı was especially known for the manufacture of iron sickles, iron scythes and iron adzes. Wrought iron therefore has deep, prehistoric roots in Zaza culture, and most appreciated were ploughshares, sickles, knives and axes.

The details of ironmaking in ancient times are contested. Iron has a melting point of 1538 °C. In a forge, the temperature can reach a maximum of 1200 °C. The forge is fired with charcoal and receives a forced supply of air with the help of bellows or a fan. The word iron is in Avestan (Old Persian) **"aêšma"** (anger). This refers to the belief that the gods gave humans iron in the form of meteors. The oldest objects in Sumer and Egypt of meteoric iron date from 4000 BC, then 2,500 years later, it was possible to extract iron from iron ore.

However, it is rarely found in pure form, but often in compounds with oxygen. An important class is iron oxide minerals such as hematite (Fe_2O_3), magnetite (Fe_3O_4), siderite ($FeCO_3$) and goethite ($FeOOH$), which are the most important ores for iron. The first iron production started in the middle of the Bronze Age, but it took several centuries before iron displaced bronze. Samples of smelted iron from

Asmar in Afghanistan, from Mesopotamia and Tall Chagar Bazar in northern Syria were made around 3000–2700 BC. The Hittites established their empire in Anatolia around 2500 BC. They seem to be the first to understand the production of iron from ore, which means a lot in their community. The Hittites began to smelt iron between 1500 and 1200 BC and the practice spread to the rest of the Near East after their empire fell in 1180 BC. The subsequent period was called the Iron Age. Artefacts of smelted iron exist in the Middle East from about 1500 BC, suggesting smelting in Anatolia. Isotopes in the metal reveal that the iron came from Eastern Anatolia.

In many towns of Eastern Anatolia, such as Tunceli, Divriği, Hekimhan and Maden, there are trigonal crystal structures up to 15 cm long of hematite minerals. According to some assertions, the Zaza people could forge these hematite crystals together with welding technology in the glow heat of the smiths' forges. Welding technology was widespread in ancient times throughout Anatolia. It occurred when an iron piece broke into two parts and was then forged at high temperatures. Even today, this welding technology is widely applicable in Eastern Anatolia.

Stone Art

In connection with Sumerian high culture, the Zaza people of the Sumerians received stone art, in particular engravings of sarcophagi. However, tombstones in the form of a large ram in the provinces of Erzincan, Tunceli, Bingöl and Elazığ first appeared in the 7th century. The old graves in the Tunceli, Erzincan, Bingöl and Elazığ regions are either on a hill or in a high place above the village, and it is possible to see religious and mythological elements as well as traditional customs and traditions on the gravestones. The gravestones in these provinces (largely sarcophagi) were worked in a plain manner. On some of the gravestones, relief figures of objects such as swords, daggers, knives, shields, forks, skewers, canes, stars, crescents and sun disks, as well as various animal and plant figures were widely used.

However, the figures were not only ornamental, but also figures indicating the gender, social position and profession of the person buried there. It is known that in the Zazas around Tunceli, Erzincan and Bingöl, a ram (man) or sheep (woman) statue is erected on the grave of the deceased according to his/her gender. Figures such as swords, daggers, knifes, shields, arrows, rams, sheep, goats indicate that the grave owner is male and has bravery. Figures such as oil lamps, scales, "saz" and similar figures belong to the clergy and figures such as needles, handles, spits, spindles and hands belong to women. Figures such as ewers and trays are expressions of generosity. Figures such as the ***Zülfikâr*** (**Zulfiqar** is the sword of Imam Ali) and the sun disks seen in some graves indicate that the person buried there is an Alevi believer.

The breakdown of the day was two 12-hour segments. In ancient times, the Zaza people used a 12-animal calendar. They also had a number system with the number 60 as the base.

Figure 37. Zaza Sarcophagus Tomb – Tunceli

A NEW EPOCH

In recent times, a lot much has happened. Today, the villages in Eastern Anatolia are largely deserted and in ruins. People have emigrated to large Turkish cities as a result of the poor economic conditions in the Zaza regions. Many villages are empty of people and no traces of the adobe (kerpiç) houses remain. Fields, meadows and forest groves have grown wild. The new inhabitants include brown bears, wild boars, wolves and birds and no man dares to go there. So far, no development in agriculture has occurred in the provinces of Erzincan, Tunceli, Bingöl and Muş. The empty Zaza villages are now leased by Turkish authorities to nomadic Kurds **(Şafaklılar)** as pastures. The Zazas are not allowed to own any forest groves, as the Turkish state owns these. Meadows and fields in forest groves are kept by the villagers. In the early 1960s, the Turkish government changed the names of all the villages that had Zazaic and Armenian names. Now, all the villages carry Turkish names.

Recent years have seen further changes. Major languages such as Turkish, mobile languages and working languages (for example, English, French and German) are taking over. The younger generation can no longer communicate with the older women, who speak only Zazaic. The Zaza society has been torn apart. Turkey is home to several different Anatolian languages, including Kurdish, Zazaic, Lazic and Syriac. All are under threat of extermination. The mother tongue of Zazaic was specific to a geographical region. A well-developed mother tongue facilitates the learning of another language and a second language as a foreign language. Expatriate Zazas are now spread all over the world and thus the language will die out. Today, Zazas live an isolated life in different countries. Moreover, mixed marriage influences some factors of a mother tongue. In addition to this, the Zaza people have no resources of their own to save the mother tongue. They lack their own written language, educated teachers, curriculum, grammars, textbooks, schoolbooks, teaching materials, teaching locals and so on.

Nearly half of the world's languages are threatened by extinction. Most of them are classified by the United Nations Cultural Organization UNESCO as proto-languages, which are deep sources of human history that are declining at an ever faster rate. Despite their deep roots, they are often small, isolated and specific to a geographical region. At the same time, the major world languages are growing. Today, the 23 largest languages are spoken by around half the world's population. In world languages, Turkish is ranked 19th and the Turkic language is ranked 5th. That big languages swallow their smaller neighbours is, in itself, a natural process. A language is considered to be extinct when no living person is mastering it. The United Nations Cultural Organization UNESCO has drawn attention to a current problem by listing Zazaic among the languages in danger of extinction. The state's responsibility to protect Indigenous peoples and their culture has not always been internationally regulated. The loss is particularly great when the language has never been documented and thus preserved for the posterity, such as Zazaic. The linguistic diversity of the world is one of our most important sources of knowledge of the varying living conditions of humans and their unique ability to adapt. A lost language is a huge loss of knowledge for humanity.

MY FAMILY TREE

My family tree starts with Cafer Ağa who is buried in 1232 in the village of Hormek in Dersim. Eleven of my ancestors' tombstones, from Cafer Ağa (died 1232) to Gülabi Ağa (died 1645), are brought to Elazığ Arkeoloji Müzesi. The year of death of my great-grandfather Muzur Teloğlu is unknown. In the wake of the famine during the Turkish-Russian war 1916-1918 my grandfather's parents Muzur Teloğlu and his wife Hava Teloğlu and grandfather's brother İbrahim Teloğlu left the village of Dılav in Erzincan to seek their subsistence in the towns on the Black Sea. They disappeared without a trace in Teloğlu history.

The great genealogy in the village of Şorik in Varto was confirmed by **Alaettin Keykubat**, the Sultan of Seljuk, in **1212**. In this pedigree, the name of Cafer Ağa, son of Melik Ağa, is mentioned. Cafer Ağa, after his father, settled in the town of **Bağın** by taking members of the tribe in Erzurum and Erzincan. Bağın is a very old Zaza city established in a wide valley of the **Kığı-Peri River**, which is located in the foothills of Tunceli Mountains and passes north of **Karakoçan** district. According to genealogy, there is now a small village in the ruins of this big town.

The genealogy was compiled from the genealogy of the Şorik village in 1543 (Hijri 950) by Ali Ağa, the son of Malhas Bey, from Cafer Ağa to Malhas Bey. Ali Rıza Sayan **("Tarihin Getirdikleri – 1978")** of Teloğulları has compiled from Mustafa Bey to Muzur Teloğlu.

Mustafa Ağa, son of Feraşat Bey, was killed in Tercan plain by the Faroğulları with a rifle in 1725 (Hijri 1137). Şişhane rifles had just come out at that time. **Teli Bey**, son of Mustafa Ağa, left Civrak village (Nazımiye) in 1788 (Hijri 1202) and came to Hoşnek subdistrict of Kığı. Teli Bey was shot in 1794 (Hijri 1209). Teli Bey's sons **Ali Teloğlu** and Hüseyin Teloğlu came to the village of **Dılav** in Selepür subdistrict in Erzincan. Since they could not get along with one of their relatives named Ali Köse, they left the village of Dılav in 1833 (Hijri 1249) and went to the village of Hoçulu in Tunceli. Unable to live in this village, they immigrated to the village of Menükler (Yarbaşı) in Kelkit. Hüseyin Teloğlu died in 1838 (Hijri 1254). The place he was fond of is called Teloğlu Plateau and Teloğlu Cemetery. Ali Teloğlu died in the village of Dılav in 1843 (Hijri 1259). Ali Teloğlu's sons İsmail-Koç Teloğlu, Haydar Teloğlu and **Muzur Teloğlu** settled in **Çamlık** hamlet near Dılav.

When the sheep flocks in the plateau were taken by Ali Köse, who was at odds before, İsmail-Koç Teloğlu, his brother Haydar Teloğlu and Muzur Teloğlu embraced their weapons and killed two people from Ali Köse, who took the sheep herd. Gülabi Ağa from Civrak village, who was an intermediary before, mediated again, and gave to the parties of Ali Köse as a blood money, from 2 of the 7 estates left by Teli Bey, the grandfather of İsmail-Koç, Haydar and Muzur. He brought these three brothers back to the village of Dılav. Haydar Teloğlu went to the sons of his uncle Hüseyin Teloğlu in the Menikler village of Kelkit. İsmail-Koç Teloğlu died in the village of Dılav in 1853 (Hijri 1269). His son Ali Teloğlu immigrated to Komga village of Başköy subdistrict of Çayırlı in 1861 (Hijri 1278). Two sons of Muzur Teloğlu, **Dursun Teloğlu** (born in 1872, death in 1963) and İbrahim Teloğlu (death year is unknown) stayed in Dılav village.

Cafer Ağa (- d. 1232 in Dersim)

↓

Kara Zeynel (- d. 1266 in Dersim)

↓

Beşir Bey (- d. 1302 in Dersim)

↓

Mümin Bey (- d. 1326 in Dersim)

↓

Zeynel Bey (- d. 1368 in Dersim)

↓

Aynal Bey (- d. 1402 in Dersim)

↓

Kara Yakup (- d. 1432 in Nazımiye)

↓

Malhas Bey (- d. 1473 in Nazımiye)

↓

Mustafa Bey (- d. 1552 in Nazımiye)

↓

Haydar Ağa (- d. 1610 in Nazımiye)

↓

Gülabi Ağa (- d. 1645 in Nazımiye)

↓

Feraşat Bey (- d. 1707 in Nazımiye)

↓

Mustafa Ağa (- d. 1725 in Tercan)

↓

Teli Bey (- d. 1794 in Erzincan)

↓

Ali Teloğlu (- d. 1843 in Erzincan)

↓

Muzur Teloğlu (- ????)/Hava Teloğlu • Süleyman/Dilber ↔ Mitibey/Fatma • Mehmet/Beser

↓

Dursun Güner (1872-1963)/Zarife Güner (1875-1946) ↔ Süleyman Balçık (1861-1937)/Elif Balçık (1864-1953)

↓

Abdullah Güner (1903-1955)/**Hatice Balçık** (1904-1951)

↓

Musa Güner

REFERENCES

Afsaruddin, Asma: *The First Muslims*. Oxford 2008.

Ahl, Augustus William: *Outline of Persian History Based on Cuneiform Inscriptions*. Nashville 2018.

Akbulut, Yılmaz: *Bingöl Tarihi*. Ankara 1998.

Akgül, Suat: *Yakın Tarihimizde Dersim İsyanları ve Gerçekler*. İstanbul 1992.

Akgün, Hüseyin: *Dersim 1938 ve Zorunlu İskân*. Ankara 2011.

Akram, Agha Ibrahim: *The Sword of Allah*. Oxford 2004.

Aksoy, Bilal: *Tarihsel Gelişim Süreci İçinde Tunceli*. Ankara 1985.

Aksoy, Ömer Asım: *Atasözler ve Deyimler*. Ankara 1965.

Akurgal, Ekrem: *Anadolu Kültür Tarihi*. Ankara 2005.

Alford, Peter; Andrews, Ezel: *Türkiye'de Etnik Gruplar*. İstanbul 1992.

Ali, Kecia: *The Lives of Muhammad*. Cambridge 2014.

Alp, Sedat: *Hitit Çağında Anadolu*. Ankara 2002.

Alvarez-Mon, Javier; Basello, Gian Pietro; Wicks, Yasmina: *The Elamite World*. Oxford 2018.

Alvarez-Mon, Javier: *The Monumental Reliefs of the Elamite Highlands*. Sydney 2019.

Arakelova, Victoria: *The Zaza People as a New Ethno-Political Factor in the Region*. Iran & the Caucasus 1999.

Arslanoğlu, İbrahim: *Şah İsmail Hatayi ve Anadolu Hatayileri*. İstanbul 1992.

Arvasi, S. Ahmet: *Doğu Anadolu Gerçeği*. Ankara 1992.

Ascalone, Enrico: *Mesopotamia. Assyrians, Sumerians, Babylonians (Dictionaries of Civilizations)*. Berkeley 2007.

Aslandoğan, Seyyid İsmail: *Kaynakları ile Alevilik*. İstanbul 2009.

Aşan, Muhammet Beşir: *Elazığ, Tunceli ve Bingöl İllerinde Türk İskân İzleri*. Ankara 1992.

Aşiretler Raporu. İstanbul 1998.

Atalay, Besim: *Bektaşilik ve Edebiyatı*. İstanbul 1991.

Balkız, Ali: *Sivas'tan Sydney'e Pir Sultan*. Ankara 1999.

Bardakçı, Cemal: *Kızılbaşlık*. İstanbul 1945.

Başbuğ, Hayri: *İki Türk Boyu Zaza ve Kurmançlar*. Ankara 1984.

Bayar, Celal: *Şark Raporu*. İstanbul 2006.

Bender, Cemşid: *Kürt Uygarlığında Alevilik*. İstanbul 1991.

Beşikçi, İsmail: *Doğuda Değişim ve Yapısal Sorunlar*. İstanbul 1969.

Bettinger, Blaine T.: *The Family Tree Guide to DNA Testing and Genetic Genealogy*. New York 2019.

Birdoğan, Nejad: *Anadolu ve Balkanlarda Alevi Yerleşmeleri*. İstanbul 1991.

Birdoğan, Nejad: *Anadolu'nun Gizli Kültürü Alevilik*. İstanbul 1990.

Birge, J. Kingsley: *Bektaşilik Tarihi*. İstanbul 1991.

Birge, J. Kingsley: *The Bektashi Order of dervishes*. London 1937.

Bojs, Karin: *My European Family*. Stockholm 2018.

Bryce, Trevor: *Life and Society in the Hittite World*. Brisbane 2002.

Bulut, Faik: *Belgelerle Dersim Raporları*. İstanbul 1992.

Cahen, Claude: *Osmanlılardan Önce Anadolu'da Türkler*. İstanbul 1979.

Cahen, Claude: *Pre-Ottoman Turkey*. New York 1968.

Crawford, Harriet E. W.: *Sumer and the Sumerians*. Cambridge 2004.

Crone, Patricia: *Meccan Trade and the Rise of Islam*. Oxford 1987.

Çınar, Erdoğan: *Aleviliğin Gizli Tarihi*. İstanbul 2015.

Dağlı, Muhtar Yahya: *Bektaşi Tomarı ve Nefesleri*. İstanbul 1935.

Dersimi, M. Nuri: *Kürdistan Tarihinde Dersim*. İstanbul 1994.

Donner, Fred M: *The Early Islamic Conquests*. Princeton 1986.

Dressler, Markus: *Die Alevitische Religion*. Würzburg 2002.

Durmuş, İlhami: *İskitler*. Ankara 1983.

Engin, İsmail; Franz, Erhard: *Aleviler/Alewiten*. Hamburg 2000.

Engin, Havva; Engin, İsmail: *Alevilik*. İstanbul 2004.

Eröz, Mehmet: *Türkiye'de Alevilik Bektaşilik*. İstanbul. 1977.

Evliya Çelebi: *Seyahatnâme 1655*. İstanbul 1986.

Faroqhi, Suraiya: *The Ottoman Empire and the World around it*. London 2007.

Fırat, M. Şerif: *Doğu İlleri ve Varto Tarihi*. Muş 1945.

Fitzpatrick, Collen; Yeiser, Andrew: *DNA and Genealogy*. Fountain Valley 2005.

Gadd, C. J: *Dynasties of Sumer and Akkad*. Bury St Edmunds 2018.

Gerger, Adnan: *Dağların Ardı Kimin Yurdu*. İstanbul 2009.

Gezik, Erdal: *Dinsel, Etnik ve Polıtık Sorunlar Bağlamında Alevi Kürtler*. İstanbul 2014.

Gezik, Erdal: *Rayberler, Pirler ve Mürşitler*. Tunceli 2012.

Gippert, Jost: *Historical Development of Zazaic*. Frankfurt 1996.

Gökalp, Ziya: *Kürt Aşiretleri Hakkında Sosyolojik Tetkikler*. İstanbul 1992.

Gökalp, Ziya: *Türkleşmek-İslâmlaşmak-Muasırlaşmak*. İstanbul 1977.

Gökalp, Ziya: *Türk Medeniyeti Tarihi*. Ankara 1976.

Gölpınarlı, Abdülbâki; Boratav, Pertev Naili: *Pir Sultan Abdal*. İstanbul 1991.

Gölpınarlı, Abdülbâki: *Oniki İmamlar*. İstanbul 1970.

Gülensoy, Tuncer: *Doğu Anadolu Osmanlıcası*. Ankara 1986.

Gülensoy, Tuncer: *Türk ve Kürt Deyim ve Atasözleri*. Ankara 1994.

Gülşan, Hasan: *Pir Hacı Bektaş Veli ve Alevi-Bektaşiliğin Esasları*. İstanbul 1975.

Gültekin, Ahmed Kerim: *Kurdish Alevism*. Leipzig 2019.

Güner, Musa: *Alevilik Tarihi*. Malmö 2025.

Güner, Musa: *History of Alevism*. Malmö 2024.

Güner, Musa: *Die Ursprünge des Zaza-Volkes basierend auf DNA*. Malmö 2022.

Güner, Musa: *Zaza People's Origin Based on DNA*. Malmö 2021.

Güner, Musa: *DNA'ya Dayalı Zazaların Kökeni*. Malmö 2021.
Güner, Musa: *Zazafolkets ursprung baserat på DNA*. Malmö 2021.
Güner, Musa: *Turkisk Grammatik*. Malmö 1997.
Gürbüz, Macit: *Kürtleşen Türkler*. İstanbul 2007.
Halm, H.: *Die islamische Gnosis: Die extreme Schia und die Alawiten*. Zurich 1982.
Hamrin-Dahl, Tina: *Aleviter i Turkiet*. Uppsala 2006.
Hezarfen, Ahmet: *Osmanlı Belgelerinde Dersim Tarihi*. İstanbul 2003.
Hormekli, Orhan Veli: *Horasan'dan Göksun'a Hormekliler*. İstanbul 2009.
İnalcık, Halil: *The Ottoman Empire*. London 1975.
İnan, Abdulkadir: *Tarihte ve Bugün Şamanizm*. Ankara 2017.
İrat, Ali Murat: *Devletin Bektaşi Hırkası/Devlet, Aleviler ve Ötekiler*. İstanbul 2008.
İzady, Mehrdad R.: *Bir El Kitabı Kürtler*. İstanbul 2004.
Jandarma Genel Komutanlığı: *Zaza Aşiretler Raporu*. İstanbul 1970.
Jones, Dorian: *Alevi Turks concerned for Alawi 'cousins' in Syria/-Globalization/DW*. Retrieved 2017.
Kafesoğlu, İbrahim: *Harezmşahlar Devleti Tarihi*. Ankara 1980.
Kafesoğlu, İbrahim: *Selçuklu Tarihi*. İstanbul 1992.
Kalafat, Yaşar: *Şeyh Sait*. İstanbul 1992.
Kaşgarlı Mahmud: *Divan-î Lugati-ût Türk*. Ankara 1985.
Kaya, Mehmed S.: *The Zaza Kurds of Turkey*. London 2018.
Kaya, Ali: *Dersim'de Dil ve Kökler*. İstanbul 2008.
Kaya, Ali: *Başlangıçtan Günümüze Dersim Tarihi*. İstanbul 2004.
Kaya, Ali: *Dersimde Kökler Doğa, İnanç, Kültür*. İstanbul 2004.
Kaya, Ali: *Deylem'den Dersime*. İstanbul 2001.
Kaya, Haydar: *Musahiplik*. İstanbul 1989.
Kaygusuz, İsmail: *İslam İmparatorlukları Tarihinde İktidar Mücadeleleri ve Aleviliğin Doğuşu*. İstanbul 2014.
Kaygusuz, İsmail: *Musahiplik*. İstanbul 1991.
Kaygusuz, İsmail: *Hünkar Hacı Bektaş Veli*. İstanbul 2001.
Kemali, Ali: *Erzincan*. İstanbul 1992.

Kennedy, Hugh: *The Prophet and the Age of the Caliphates*. Harlow 2004.

Kershasp, P.: *Studies in Ancient Persian History*. London 2013.

Khalidi, Tarif: *Images of Muhammad*. New York 2009.

King, Leonard W.: *A History of Sumer and Akkad*. London 2018.

Kingsley, Patrick: *Turkey's Alevis, a Muslim Minority, Fear a Policy of Denying Their Existence*. New York 2017.

Kinzer, Stephen: *Crescent and Star. Turkey between Two Worlds*. New York 2003.

Koca, Turgut: *Bektaşi Nefesleri ve Şairleri*. İstanbul 1990.

Kocadağ, Burhan: *Lolan Aşireti ve Yakın Çevre Tarihi*. İstanbul 2007.

Kocadağ, Burhan: *Doğuda Aşiretler, Kürtler, Aleviler*. İstanbul 2004.

Kocadağ, Burhan: *Alevi-Bektaşi Tarihi*. İstanbul 1998.

Konukçu, Enver: *Köroğlu'na Kadar Bingöl*. Ankara 1987.

Köhler, Wilhelm: *Evliya Çelebi Seyahatnamesinde Bitlis ve Halkı*. İstanbul 1989.

Kramer, Samuel Noah: *History Begins at Sumer*. Philadelphia 1988.

Kramer, Samuel Noah: *Tarih Sümer'de Başlar*. Ankara 1995.

Kudat, Ayşe Sertel: *Kirvelik*. Ankara 1974.

Küçük, Hülya: *The Roles of the Bektashis in Turkey's National Struggle*. Leiden 2002.

Lapidus, Ira M.: *A History of Islamic Societies*. Cambridge 2002.

Lerch, Peter: *Forschungen über die Kurden und die Iranischen Nord-chaldäer*. St. Petersburg 1857.

Lloyd, Seton: *The Archaeology of Mesopotamia*. London 1978.

Maalouf, Amin: *Korstågen enligt araberna*. Lund 2001.

Madelung, Wilfred: *The Succession to Muhammad*. Cambridge 1997.

Mann, Oskar; Hadank, Karl: *Die Mundarten der Zaza hauptsächlich aus Siverek und Kor*. Leipsig 1932.

McDowall, David: *Modern Kürt Tarihi*. Ankara 2004.

Melikoff, Irene: *Hacı Bektaş: Efsaneden Gerçeğe*. İstanbul 1998.

Molineux, L.: *A Journey in Dersim*. London 1914.

Momen, Moojan: *An Introduction to Shii Islam*. New Haven 1985.

Morimoto, Kazuo: *Sayyids and Sharifs in Muslims Societies*. London 2012.

Mutin, Benjamin; Lamberg-Karlovsky, C. C.: *The Proto-Elamite Settlements and Its Neighbors*. Cambridge 2014.

Nakash, Yitzhak: *The Shiis of Irak*. Princeton 2003.

Nehar, Beran: *Zazalar Tarihi*. İstanbul 2020.

Nemet-Nejat, Murat; Rhea, Karen: *Daily Life in Ancient Mesopotamia*. London 1998.

Nuri, İhsan: *Kürtlerin Kökeni*. İstanbul 1977.

Ocak, Ahmet Yaşar: *Alevi Bektaşi İnanışlarının İslam Öncesi Temelleri*. İstanbul 2000.

Olsson, Tord; Özdalga, Elizabeth; Raudvere, Catharina: *Alevi Identity*. İstanbul 1998.

Orhunlu, Cengiz: *Osmanlı İmparatorluğu'nda Aşiretlerin İskanı*. İstanbul 1987.

Oytan, M. Tevfik: *Bektaşiliğin İçyüzü*. İstanbul 1994.

Önder, Ali Tayyar: *Türkiye'nin Etnik Yapısı*. Ankara 2008.

Öz, Baki: *Osmanlı'da Alevi Ayaklanmaları*. İstanbul 1992.

Özcan, Mesut: *Dersim (Zaza) Atasözleri*. İstanbul 1992.

Özdemir, Ali Rıza: *ZAZALAR ve TÜRKLÜK*. Ankara 2016.

Özdemir, Ali Rıza: *ZAZALAR KÜRTLER ALEVİLER*. Ankara 2015.

Özkırımlı, Atilla: *Alevilik - Bektaşilik Edebiyatı*. İstanbul 1985.

Özkul, Derya; Markussen, Hege: *The Alevis in Modern Turkey and the Diaspora*. Edinburgh 2022.

Öztelli, Cahit: *Pir Sultan Abdal*. İstanbul 1989.

Pamukçu, Abubekir: *Dersim Zaza Ayaklanmasının Tarihsel Kökenleri*. İstanbul 1992.

Pomery, Chris; Jones, Steve. *DNA and Family History*. Toronto 2004.

Potts, Daniel T.: *The Archaeology of Elam*. Cambridge 1999.

Refik, Ahmet: *Anadolu'da Türk Aşiretleri*. İstanbul 1989.

Reich, David: *Who We Are and How We Got Here*. Cambridge 2019.

Rişvanoğlu, Mahmut: *Doğu Aşiretleri ve Emperyalizm*. İstanbul 1992.

Rişvanoğlu, Mahmut: *Saklanan Gerçek: Kurmançlar ve Zazaların Kimliği*. Ankara 1994.

Sait, Baha: *Türkiye'de Alevi-Bektaşi, Ahi ve Nusayrî Zümreleri*. Ankara 2000.

San, M. Salih: *Doğu Anadolu ve Muş'un İzahlı Kronolojik Tarihi*. Ankara 1982.

Sani, Nader: *Muhammeds alla hustrur*. Nyköping 1992.

Saraç, Necdet: *Alevilerin Siyasal Tarihi*. İstanbul 2010.

Say, Yağmur: *Anadolu Alevilerinin Tarihi*. İstanbul 2007.

Sayan, Ali Rıza: *Tarihin Getirdikleri*. İstanbul 1978.

Selcan, Zılfi: *Zaza Dilinin Gelişimi*. Berlin 2004.

Sevgen, Nazmi: *Zazalar ve Kızılbaşlar*. Ankara 1999.

Shaban Muhammad A: *Islamic History*. Cambridge 1976.

Shindeldecker, John: *Turkish Alevis Today*. İstanbul 1996.

Sjölund, Peter: *Släktforska med DNA*. Stockholm 2015.

Sökefeld, Martin: *Cem in Deutschland*. Frankfurt 2005.

Sümer, Faruk: *Doğu Anadolu'da Türk Beylikleri*. Ankara 1993.

Sümer, Faruk; Sevim, Ali: *Malazgirt Savaşı*. Ankara 1993.

Şahhüseyinoğlu, Nedim: *Alevi Örgütlerinin Tarihsel Süreci*. Ankara 2001.

Şener, Cemal: *Türkiye'de Yaşayan Etnik ve Dinsel Gruplar*. İstanbul 2006.

Şener, Cemal: *Osmanlı Belgelerinde Dersim Tarihi*. İstanbul 2006.

Şener, Cemal; Miyase, İlknur: *Şeriat ve Alevilik: Kırklar Meclisi'nden Günümüze Alevi Örgütlenmesi*. İstanbul 1995.

Şener, Cemal: *Alevi Törenleri*. İstanbul 1991.

Şeref Han: *Şerefnâme 1597; Kürt Tarihi*. İstanbul 1998.

Şapolyo, E. Behnan: *Mezhepler ve Tarikatlar Tarihi*. İstanbul 1964.

Tahran, Taner: *Eski Çağda Kimmerler Problemi*. Ankara 1979.

Taneri, Aydın: *Harezmşahlar*. Ankara 1993.

Tankut, Hasan Reşit: *Zazalar Üzerinde Sosyolojik Tetkikler*. İstanbul 2000.

Tankut, Hasan Reşit: *Aleviliğin Menşei*. İçel 1938.

Teker, Mehmet: *Kürt Tarihinde Hormek (Alhas) Aşireti*. Ankara 2006.

Timuroğlu, Vecihi: *Dersim Tarihi*. Ankara 1991.

Toker, Metin: *Şeyh Sait ve İsyanı*. Ankara 2015.

Tosun, Halis: *Alevi Kimliğiyle Yaşamak*. İstanbul 2002.

Tosun, Mebrure; Yalvaç, Kadriye: *Sümer Dili ve Grameri*. Ankara 1981.

Turan, Osman: *Selçuklular Tarhihi ve Türk-İslam Medeniyeti*. İstanbul 2008.

Türk, Hüseyin: *Nusayrilik*. İstanbul 2000.

Türkay, Cevdet: *Osmanlı İmparatorluğunda Oymak-Aşiret ve Cemaatlar*. İstanbul 1979.

Türkdoğan, Orhan: *Zazalar ve Kürtler*. İstanbul 2010.

Türkdoğan, Orhan: *Alevi-Bektaşi Kimliği: Sosyal-Antropolojik Araştırma*. İstanbul 1998.

Uluçay, Ömer: *Alevilikte Cem Nefesleri*. Adana 1996.

Ulusoy, Celalettin: *Hünkâr Hacı Bektaş Veli*. Hacıbektaş 1986.

Vallat, François: *The History of Elam. The Circle of Ancient Iranian Studies*. Caen 2010.

Vorhoff, Karin: *Zwischen Glaube, Nation und neuer Gemeinschaft: Alevitische Identität in der Türkei der Gegenwart*. Berlin 1995.

Werner, Eberhard: *Rivers and Mountains*. Nürberg 2017.

Woodhead, Christine: *The Ottoman World*. London 2011.

Yaman, Mehmet: *Alevilik İnanç, Edep, Erkân*. İstanbul 2013.

Yaman, Mehmet: *Alevilikte Cem*. İstanbul 2003.

Yaman, Ali: *Alevilikte Dedelik Kurumu ve İşlevleri*. İstanbul 1996.

Yaman, Ali; Aykan, Erdemir: *Alevism-Bektashism*. London 2006.

Yıldırım, Ali: *Başlangıcından Bugüne Alevi-Bektaşi Deyişleri*. Ankara 1997.

Yılmaz, Abdurrahman: *Tahtacılarda Gelenekler*. Ankara 1948.

Yörükan, Yusuf Ziya: *Anadolu'da Aleviler ve Tahtacılar*. Ankara 1998.

Zaman, Muhammad Qasim: *The Ulama in Contemporary Islam*. Princeton 2002.

Other works by Musa Güner

Zaza Tarihi, *Gece Kitaplığı 2025*

Alevilik Tarihi, *Gece Kitaplığı 2025*

History of Alevism, *LIT Verlag 2024*

Die Ursprünge des Zaza-Volkes basierend auf DNA, *LIT Verlag 2022*

Zaza People's Origin Based on DNA, *Nobel Akademik Yayıncılık 2021*

DNA'ya Dayalı Zazaların Kökeni, *Festival Yayıncılık 2021*

Zazafolkets ursprung baserat på DNA, *Förlag Güner 2021*

Svensk-Turkisk Parlör, *Förlag Güner 2015*

Svensk-Turkisk / Turkisk-Svensk Fickordbok, *Förlag Güner 2014*

Turkisk-Svensk / Svensk-Turkisk Ordbok, *Förlag Güner 2011*

Turkisk-Svenskt Lexikon, *Förlag Güner 2009*

Svensk-Turkiskt Lexikon, *Förlag Güner 2004*

Türkçe-İsveççe / İsveççe-Türkçe Sözlük, *Förlag Güner 1998*

Turkisk Grammatik, *Förlag Güner 1997*

Svensk-Turkisk / Turkisk-Svensk Handordbok, *Bokförlaget Corona AB 1992*

Turkisk-Svensk Ordbok, *Studentlitteratur AB 1985*

Svensk-Turkisk Ordbok, *Studentlitteratur AB 1984*

Geochemistry and Tectonic Significance of Altered Basaltoids from the Pontic Ranges, Turkey, *Geologische Rundschau 1983*

A Paleomagnetic Study of Some Basaltoids and Ores from the Pontic Ranges, Turkey, *Tectonophysics 1982*

Turkisk-Svensk Ordbok, *Förlag Güner 1981*

Türkçe-İsveççe Sözlük, *Studentlitteratur AB 1981*

Geology and Massive Sulphide Ore of the Küre Area, the Pontic Ranges, Turkey, *MTA 1980*

A Note on Chrome Spinels and Ore Minerals from the Küre Area, the Pontic Ranges, Turkey, *MTA 1979*

İsveççe-Türkçe Sözlük, *Studentlitteratur AB 1977*

Svensk-Turkisk Ordbok, *Förlag Güner 1976*

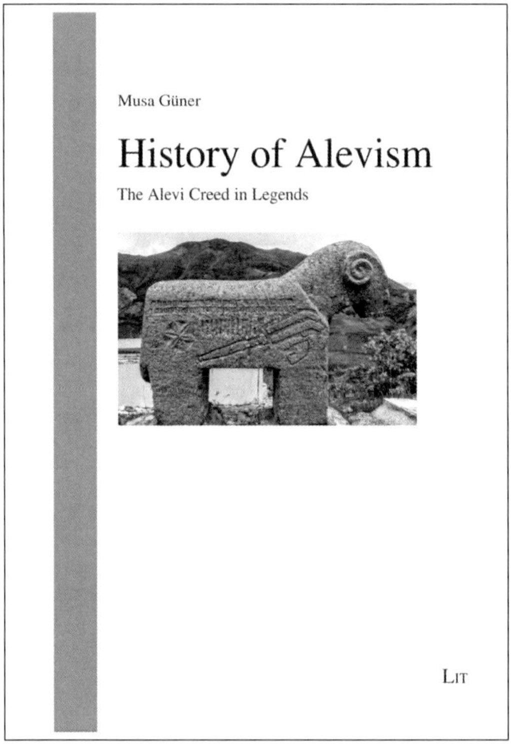
Musa Güner
History of Alevism
The Alevi Creed in Legends
LIT

Kulturwissenschaft / Cultural Studies / Estudios Culturales / Études Culturelles

Musa Güner

Die Ursprünge des Zaza-Volkes basierend auf DNA

Die Geschichte des unbekannten Zaza-Volkes

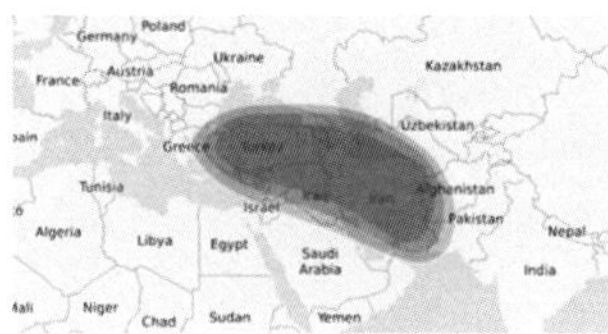

Kulturwissenschaft Bd. 68

LIT

Musa Güner

Die Ursprünge des Zaza-Volkes basierend auf DNA

Die Geschichte des unbekannten Zaza-Volkes

Der Band widmet sich der Geschichte der Zaza. Es wird geschätzt, dass weltweit insgesamt 1.500.000 (in Deutschland ca.170.000) Zaza leben. Die Zaza sind die drittgrößte ethnische Gruppe in der Türkei. DNA-Tests belegen die Verwandtschaft der Zaza-Stämme miteinander. Untersuchungen über DNA, Sprachen, Migrationen, religiöse Umgebungen, Lebensgewohnheiten, Landwirtschaft und kulturelle Aktivitäten tragen zum Wissen über die Ursprünge des Zaza-Volkes bei und werden thematisiert.

Bd. 68, 2022, 96 S., 19,90 €, br., ISBN 978-3-643-25042-1

LIT Verlag Berlin – Münster – Wien – Zürich – London

Auslieferung Deutschland / Österreich / Schweiz: siehe Impressumsseite